The Effects of Traffic Structure on Application and Network Performance

T0185100

Jay Aikat · Kevin Jeffay · F. Donelson Smith

The Effects of Traffic Structure on Application and Network Performance

 Springer

Jay Aikat
Department of Computer Science
University of North Carolina at Chapel Hill
Chapel Hill, NC, USA

Kevin Jeffay
Department of Computer Science
University of North Carolina at Chapel Hill
Chapel Hill, NC, USA

F. Donelson Smith
Department of Computer Science
University of North Carolina at Chapel Hill
Chapel Hill, NC, USA

ISBN 978-1-4939-0038-1   ISBN 978-1-4614-1848-1 (eBook)
DOI 10.1007/978-1-4614-1848-1
Springer New York Heidelberg Dordrecht London

Printed on acid-free paper

Springer is part of Springer Science+Business Media (www.springer.com)

Contents

Chapter 1
Introduction

When one discovers a fact about nature, it is a contribution per se, no matter how small. Since anyone can create something new [in a synthetic field like Computer Science], that alone does not establish a contribution. Rather, one must show that the creation is better. Accordingly, research in computer science and engineering is largely devoted to establishing the "better" property.

Fred Brooks [3]

Over the past three decades, the Internet's rapid growth has spurred explosive development of new applications such as mobile computing, digital music, and online video and gaming. The performance of these applications depends on the performance of various protocols and mechanisms enabling Internet functions. For 30 years now, TCP (Transmission Control Protocol) and IP (Internet Protocol) have been the dominant communication protocols, and they have fortuitously evolved despite the Internet's multifold growth. To improve the Internet's performance, networking researchers constantly develop new protocols and innovations.

These protocols must be tested before they can be deployed on the Internet. In most fields, there are agreed-upon standards to test such new inventions and improvements. For example, if Intel develops a new processor, several benchmarks test the new processor to demonstrate that it performs better than an existing one. However, computer networking, as a nascent field with explosive growth, still lacks such standards for protocol evaluation. Establishing such standards remains a challenging research endeavor in networking, and it forms the central motivation for the research presented in this book. By generating different kinds of network traffic within the laboratory testbed, showing how and why the use of different models of application workload and network path characteristics during traffic generation affect the outcome of experimentation, we have asked and answered some fundamental

J. Aikat et al., *The Effects of Traffic Structure on Application and Network Performance*,
DOI 10.1007/978-1-4614-1848-1_1, © Springer Science+Business Media New York 2013

questions about experimental methodology in networking research. We plan to use the lessons learned from this study to motivate further discussions and concrete steps in the networking research community toward establishing better practices in experimental methods for networking research.

Networking researchers have long used experimental networks and distributed systems for evaluating new networking technologies. Indeed, experimentation, either via software simulation using simulators such as the Network Simulator [4], or via hardware emulation using laboratory testbeds, has been the primary means for evaluating existing and newly proposed protocols and algorithms for improving the Internet. Hence, improving the Internet involves constantly improving the process of experimentation to produce reliable and reproducible results for empirical evaluations. This requires research into methodology. This research study is a step in that direction. Experimental methodology has many components. The study presented in this book is methodological in nature, exploring one major component – traffic generation in network experimentation.

1.1   Traffic Generation

One of the most complex components of empirical evaluations is modeling and generating realistic Internet traffic. The mix of the ever changing and varied applications that constitute actual Internet traffic makes this a daunting task. Moreover, Internet traffic is different when sampled at different times and in different parts of the globe.

Networking researchers have grappled with this problem by taking snapshots of Internet traffic at different times and at various points in the network, and modeling the same for generating traffic in the lab. The generally held belief is that the more realistic the traffic used, the more reliable are the results of the empirical evaluations using that traffic. Practice, however, does not adhere to this principle. So, although laboratory testbeds and methods for simulations have evolved over the years, the question about what constitutes essential components for modeling realistic traffic remains open for debate.

For example, networking researchers agree that realistic traffic generation for empirical research is best accomplished by capturing traffic on a production link and then using source-level models to generate this traffic in the laboratory or simulator. Source-level models capture the application exchanges and application behavior on the ends (sources) of the TCP connections. But how do you go from the original captured traffic to an acceptable source-level model? Which of the several measures derived from the traffic sources should you model in your workload for your experiments? Would your modeling choices for traffic generation impact the outcome of your experiments? If yes, how significant would the impact be? These remain open questions.

Let's consider an example. Say you developed a new high-speed variant of TCP; let's call it TCP-X. To show that TCP-X is indeed better than the existing variants of TCP on the Internet today, you would need to run some experiments either in a laboratory setting or using a simulator. You would not wish to run your experiments directly on the Internet as that will reduce control, and you could not repeat your experiments under the same conditions. Moreover, injecting traffic using untested protocols with possible bugs can cause breakdown of network services. So, as part of the experiment using a laboratory testbed or simulator, you would generate traffic between sets of endpoints (traffic generators) that use either TCP-X or the other TCP variant against which you are testing your new protocol. For your experiments, you need to generate realistic traffic. So you collect network traffic on a production link. Since you are testing the performance of transport protocols, you decide to use application workload models (source-level models) for generating traffic. That is, you generate traffic in your experiments by driving network stacks with the application models derived from your empirical measurements and you use the applicable TCP protocol on the endpoints. You choose this approach because traffic generated in this *closed-loop* manner fully preserves the fundamental feedback loop between the network endpoints and network characteristics. This is essential for testing transport-level properties.

Now, having made all these decisions on experimental design, how would you use the captured traffic from that production link to drive the network stacks on these endpoints or traffic generators? That is, given the empirical measurements of the traffic you captured, which of those measurements will you use to create your application workload models for generating traffic in your experiments, and why?

Let us consider some possible choices in modeling the workload you captured. You have the packet header trace which can be used to derive a lot of information on every TCP connection constituting that traffic. Do you send all the measured bytes for a given TCP connection as one large data unit in each direction? If yes, do you send them concurrently in both directions, or do you simulate a request-response behavior between a client and server, and thus send all the data in one direction first, then send all the data in the other direction? Say, you use one of these two methodologies to generate a persistent HTTP connection that originally had request-response, request-response, and so on, with 25 such request-response pairs sending data back and forth between client and server. Have you somehow distorted this connection by generating it all as one large data unit in each direction? If you have, does it matter? If it matters, when does it matter? That is, what performance metrics (output results that you use to show that TCP-X is better than other TCP variants) are affected favorably or adversely by such distortion of the connections generated during the experiment?

In the above scenario of generating a persistent HTTP connection, the original connection could be represented in two dimensions – size and time. The size component is the data being sent all at one time or in small chunks back and forth as measured in the original connection. But while the size component of a connection seems obviously necessary for representing the connection for traffic generation, what role does the time component play, and how does it affect the performance metrics in your experiment? In fact, there is more than one time component in any given connection. There are the times between packets sent on the network, times between a request and its corresponding response, and the user-generated thinktimes which are the times between consecutive request-response exchanges within a connection. Which of these do you model, and how would your experimental results be affected by your choices?

So far, we have only discussed the application workload in both the size and time components. But for realistic traffic generation, we must consider that the endpoints or traffic generators that generate this application workload are also influenced by the network conditions. This brings in another time component of traffic generation – the connection round trip time (RTT). What is the best method of RTT emulation? Is one method better than another, and why? For example, if you determine that the mean RTT of all connections in your measured traffic was 80 milliseconds, could you use this as the default RTT for every connection in your experiment? How would this choice for RTT emulation influence the performance metrics you study in your experiment? What if, instead, you measured the connection RTT for every connection on that original link, and faithfully assigned each generated connection its measured RTT during your experiment? What is the benefit of such a choice in generating traffic?

How will your choice of parameters for application workload modeling, and your choice of model for emulating network path characteristics like RTT emulation, affect the outcome of your experiments? That is, how will these choices affect whether your TCP-X shows *better* results than some popular variant of TCP for the metrics you are using in this evaluation? Say the results showed that TCP-X is indeed *better* for certain metrics of performance than other TCP variants. Would you then be able to use the results from such an experiment with confidence to deploy TCP-X on the Internet? Why or why not?

These are the kinds of questions that motivated this study. We strive to advance such discussion and the exploration of experimental methodologies in networking research. We developed a spectrum of empirically-derived, realistic models for generating TCP traffic, and different models for emulating RTT, in the laboratory. We conducted experiments using this spectrum of application workload models we call TCP connection structures and round trip time (RTT) emulation methods – all inspired by models used in leading publications. Our goal was to explore how generating the same empirically-derived traffic using different connection structures and different RTT emulation methods alters key characteristics of traffic in the network, thus affecting the user perceived performance metrics of connection durations and response times as well as network centric metrics of active connections and router queue lengths.

1.2 The Tmix Traffic Generation System

This study is based on the foundation laid by the Tmix traffic generation system developed by Hernandez-Campos et al. [2, 7]. In that work, the authors presented a new methodology for generating network traffic using source-level modeling in testbed experiments and software simulations. They developed a new source-level model of network traffic, the *a-b-t* model (we call this the *a-t-b-t* model in our study), for describing in a generic and intuitive manner the behavior of the applications driving the TCP connections in network traffic. Hernandez-Campos et al. made the following major contribution: they showed that given a packet header trace collected at any Internet link, their Tmix traffic generation system reproduced the application-level behavior as well as network-level parameters, like RTT and window size. The statistical properties of this synthetically generated traffic matched very closely with those of the original traffic captured on the Internet link. We verify this demonstration as part of our calibration experiments, and hence use their model as the control set for our experiments. We use the Tmix traffic generation system in our research, and adopt their terminology to explain our models for application workloads. Hence, in this section we explain their terminology.

The Tmix traffic generator is an empirically-based approach to workload generation. Starting from a trace of TCP/IP headers collected on a production network, they constructed a model for all the TCP connections observed in the network without knowledge of the underlying applications. The model, a set of *a-t-b-t* connection vectors, can be used in the workload generator Tmix to generate the connections and reproduce the application-level behaviors observed on the original network. That work also identifies a fundamental dichotomy in source-level behavior between connections that exchange data sequentially and those that exchange data concurrently.

In Tmix each connection found in a trace of TCP/IP headers from a production network link is analyzed to produce a "*connection vector*" representation. The connection vector includes the connection's start time relative to the beginning of the trace and a series of request-response exchanges found by their analysis tool. Each request-response exchange (called an "*epoch*") is described by a 4-tuple consisting of the request size (called the "a" unit size), the response size (called the "b" unit size) and two latency values (called the "t" values) for the time between a request and its response and for the time between successive request-response exchanges. Unidirectional transfers have only an 'a' or 'b' value depending on the direction of transfer.

Our definitions and models for traffic generation in this study derive heavily from this work [2]. Hence, a high level summary of the Tmix analysis and generation framework is given in Fig. 1.1. The first step in this process is to capture a trace of TCP/IP headers on any production link. This trace is then processed to produce a set of connection vectors such that each TCP connection in the trace is now defined by a unique connection vector. The Tmix traffic generation tool takes as input this set of connection vectors and replays these connections to produce traffic on the link such that its statistical properties match those of the traffic that was originally captured.

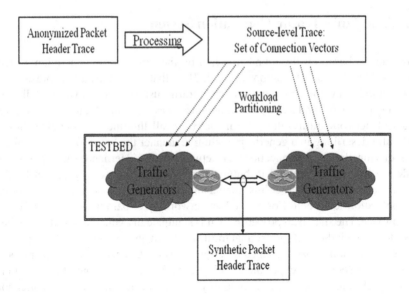

Fig. 1.1 Traffic Generation

1.3   Modeling TCP Connection Structure

In this study, we used Tmix's *a-t-b-t* model as a control for all our connection
structure models. We define connection structure for a TCP connection as modeled
in two dimensions – size and time. The size dimension defines the total number of
bytes transferred by the connection in both directions. The time dimension models
the internal dynamics of a connection consisting of any synchronization and latencies introduced by exchanges of application-level protocol data units, typically in a
request-response pattern as in a client-service model of communication. The time
dimension includes all the latencies related to synchronization between requests and
responses (modeling *epochs*), the elapsed time between a request and its response
(server latency, or intra-epoch latency) or between requests (client latency also
called user thinktime, or inter-epoch latency). In connections that send data concurrently in both directions, the time dimensions represent the *quiet* periods between
transmissions of application data units in either direction.

We represent connection structures in this study by starting with a simple model,
based on Harpoon [5], defining the connection structure in the size dimension alone.
Consider a connection that transfers a total of X bytes in one direction between
endpoints and Y bytes in the opposite direction over the duration of the connection.
Harpoon generates two separate connections to model each original connection
with a unidirectional transfer of all the bytes in each direction as a single block in
each of the two generated connections. We modified this concept to use a single
TCP connection for each original connection, but with two different methods of
synchronizing the bidirectional data transfers.

In both methods, all the bytes flowing in one direction are sent as one large block without internal gaps or latencies. In one method the two blocks are sent concurrently in both directions while in the other method the two blocks are sent sequentially as a request-response exchange. We call the first method the block-concurrent (*blk-conc*) model and the second method the block-sequential (*blk-seq*) model.

The three ways of representing connection structure described so far (Harpoon, block-concurrent, and block-sequential) are all based solely on the size dimension of connections. To introduce the time dimension, we turn to the representations exemplified by the Swing [6] and Tmix [2] traffic generators. Using Tmix's *a-t-b-t* framework, we can describe several variations for representing connection structures. First, we retain the set of epochs representing the request-response exchanges along with the *a* and *b* values for each epoch but without any of the *t* values. This representation that we call the *a-b* model includes the time dimension only in the implied synchronization between a request and its response.

Next, we define the *a-t-b* model in which the *t* represents the full latency between a request and its response thus implicitly representing any server processing time. The full representation of a connection, the *a-t-b-t* model, adds the latency between successive requests and thus any client processing or user *think* times. Thus we start with only the size dimension to model a TCP connection, and add in the time dimension creating six slightly different models for any TCP connection. These are the *Harpoon*, *blk-conc*, *blk-seq*, *a-b*, *a-t-b*, and the *a-t-b-t* models.

1.4 Emulating Network Path Characteristics

In addition to experimenting with six models for representing connection structure for a TCP connection, we ran experiments using seven different methods of emulating round trip times (RTTs) in our experiments. All of these have either been used in, or are inspired by, previously published works. For one extreme we first tried emulating no RTT latency (*nodelay*) beyond that inherent in the laboratory network used in the experiments which is typically 1 millisecond or less (reasonable for studying local area networks but obviously wrong for wide-area network emulation). At the other extreme, we used the Tmix capability (called *usernet*) to emulate the specific minimum RTT measured for each connection from the empirical analysis of the originally captured TCP/IP header traces.

In addition to the *nodelay* and *usernet* RTT models, we developed five more models as follows. First, we emulated a single non-zero value for all connections, using either the mean or median of the RTTs found by analyzing all the several million connections in the TCP/IP header traces. The "*nodelay*", mean and median RTT cases all represent one method of assigning a single value to all connections in the hour long experiment. This method of assigning connection RTTs effectively emulates a single end-to-end network path for all the connections in the experiment. We then created models emulating *n* network paths by assigning a specific round trip time delay to each of the *n* end-to-end paths where *n* was 10 in one model and 30 in another.

The network used in this study has a maximum of 30 pairs of traffic generator machines. In one case, we assigned a unique emulated RTT to a path shared among three pairs (a total of 10 end-to-end path RTTs). The values chosen for this case were the values recommended for the TMRG common TCP evaluation suite [1]. In a second variation, we assigned a unique RTT value for each of the 30 end-to-end paths between the 30 pairs of traffic generator machines. In this case, we used a discrete approximation method to approximate the empirical RTT distribution found from analysis of the traces.

Finally, we ran experiments using an RTT emulation method where a value was assigned per connection to each of the several million connections in the experiment. In one case, we assigned to each connection a value randomly sampled from a uniform distribution of RTT values. In the other case, we used the Tmix method of assigning the specific minimum RTT for each connection as observed in the originally captured TCP/IP header trace. Thus we used seven different RTT emulation methods, three assigning values on a per-experiment basis (one RTT value for all connections), two assigning values on a per-path basis and two more on a per-connection basis.

We ran experiments with the full cross product of six connection structures and seven RTT emulations, and we report those results in Chapters 5 and 6.

1.5   Changing the Network Environment

We ran every experiment described above in two distinct environments in the network. First, we set the link between the two routers in the *unconstrained* network mode where the link capacity was unchanged at 1Gbps. Next, we set the link in *constrained* network mode where the link bandwidth was limited such that the link capacity was 105% of the traffic traversing the link. Setting constraints on this link enabled us to study the queue dynamics for the outgoing queue at the router before this link. See Chapter 4 for details on network setup and topology.

We first ran all calibration experiments in *unconstrained* mode. Then we ran experiments using the different connection structure models and RTT emulations in both *unconstrained* and *constrained* modes to study the effect of changing the network environment on network performance. We detail the results from these experiments and discuss the effect of the network environment on the outcome of experiments in Chapters 5 and 6. Chapter 5 presents results for a complete set of experiments run in both network environments. Chapter 6 presents interesting, additional results for experiments run in one or both network environments.

1.6   Using Two Input Traces

To ensure robustness of our results, we ran all our experiments using two input traces collected at two diverse locations on the Internet. The first one from UNC was taken on the border link connecting the campus to the Internet service provider network.

The second trace was taken at an aggregation switch for four internal networks, connecting one of IBM Corporation's largest development sites to the Internet. The UNC campus trace was a 1-hour trace on a weekday during the school year. The IBM trace was also a 1-hour trace which was representative of typical peak workday traffic on this corporate network. The UNC trace has almost 4.7 million connections with an average load of 471 Mbps in one direction and 202 Mbps in the other. The IBM trace has about 2.8 million connections with an average load of 404 Mbps in one direction and 366 Mbps in the other.

1.7 Modeling Receiver Window Sizes

For all the experiments exploring connection structure models and RTT emulation methods, we used Tmix's model for assigning window sizes to the two ends of every TCP connection. Each side of every connection was assigned the maximum receiver window size exactly as measured through the analysis of the original trace. Hence, even when we modeled the simplest connection structures like block-concurrent, we provided some inherent sophistication to the overall traffic modeling by the assignment of measured receiver window sizes. Our decision here was based on the idea that a system is best studied when adjusting one tunable knob at a time. Hence we kept the window size for connections in these experiments the same as empirically observed in the original header trace.

Besides the full suite of experiments using different connection structure models and different RTT emulation methods, we ran experiments where the maximum receiver window sizes were fixed for all connections as 8 KB, 16 KB, or 64 KB, using only the control set combination of the *a-t-b-t* connection structure and *user-net* RTT models. Results for these experiments are reported in Chapter 6.

1.8 Hypotheses

This study is based on the following hypotheses:

> The structure of application workload models (TCP connection structure) and the characteristics of the network path through the emulation of Round-Trip-Time (RTT) models, significantly impact the outcome of experiments. Such impact can be quantitatively demonstrated through measurement of performance metrics both by the user-perceived performance metrics of application behavior as well as network-centric performance metrics at the routers and links in the network.

And through extensive laboratory experimentation and analyses, we show how specific modeling choices in traffic generation affect the outcome of the experiments in which they are used. The outcome of any experimental evaluation depends heavily on the input to the system – this is the *garbage-in garbage-out* concept. Based on the detailed study of the behavior of standard TCP and its high-speed variants by many leading researchers as well as preliminary laboratory experiments, our initial hypotheses was that the application workload and network path characteristics applied as input to the research network testbed system heavily impact the resulting application and network behavior. Within the realm of empirically-derived traffic generation, our goal was to differentiate among different aspects of emulating application workloads and network path characteristics, and show how they affect performance metrics both at the network-level and the application-level.

1.9   Summary of Conclusions and Contributions

Through extensive experimentation using the Tmix traffic generation system as the basis for running experiments on a laboratory testbed, we arrive at the following conclusions.

> In an unconstrained network, regardless of the application workload model used, or the input traffic used, round trip time had a significant effect on user perceived performance measures of connection duration and response times, but only up to a maximum of 1 second of the distribution for these metrics.

With no constraint on the link, we found that different round trip time models used in traffic generation affect experimental outcomes differently. As expected, we found that different RTT models resulted in different distributions of connection duration and response times. These differences, however, were significant only up to about 500 milliseconds, or a maximum of 1 second of the distribution for these metrics. Beyond that, the RTT model has little effect on these metrics.

> RTT model had no impact on the number of active connections (measured in 1 second intervals) in the network.

The number of active connections in the network is a second order measure of performance and a key metric for many router protocol evaluations. It is directly affected by the durations of connections in the network. Since the choice of RTT model affects the distribution of connection durations only up to 1 second of the distribution, and since we compute a connection to be active in one second intervals, the RTT models emulated in our experiments do not affect the number of active connections in the network.

> In a constrained environment, the smaller the median of the distribution of connection RTTs, the heavier the resulting queue distribution was at the router.

When the router-to-router link is *constrained*, the different round trip time models used in generating traffic alter the queuing dynamics at the router before the *constrained* link to slightly different degrees. In such a *constrained* mode, some RTT models cause larger queuing delays than others. For example, let us compare two experiments – one in which we used the *usernet* RTT model which has thousands of connections with small RTTs (median RTT for this distribution was 36 milliseconds), and the second in which we use one value of 80 milliseconds as the

RTT for all connections in the experiment (80 ms was the mean of the *usernet* RTT distribution). We found that for a given connection structure model, using the *usernet* RTT model resulted in the heavier queue length distribution because for a large number of connections, their RTTs were smaller than the 80 ms RTT assigned to all connections in the *meanRTT* model. The experiments using *meanRTT* resulted in relatively lighter queue distributions.

In a constrained environment, there were no differences in connection durations or response times due to different RTT models for the block and a-b connection structure models.

In an *unconstrained* environment, we observed clear differences in connection durations and response times due to different RTT models for the block and a-b connection structure models. However, in the *constrained* mode, the block and a-b models resulted in very heavy queue distributions. This caused long enough queuing delays that almost completely masked the differences in distribution of connection durations and response times among the three connection structure models. The distribution of these metrics, however, had shifted heavily in the *constrained* mode as compared to their corresponding *unconstrained* experiments. The only connection structure for which RTT models still made a difference on these metrics in the *constrained* mode was the *a-t-b-t* model. This is because the *a-t-b-t* model does not create as heavy queues as the other connection structure models. Hence when using the *a-t-b-t* model, the differences in connection duration and response times up to 1 second of the distributions were still observed in the *constrained* mode.

Randomly assigning the same empirically derived round trip times to connections, using the discrete-approximation RTT model, is almost as effective, on an aggregate level in the unconstrained mode, as assigning each connection its originally measured RTT using the usernet model.

We developed an approximation of the empirical RTT distribution from the *usernet* model; we called this the *discrete approximation* or the *DA* RTT model. We found that the *DA* model for RTT emulation yields results for all metrics very similar to the *usernet* model in the unconstrained mode, as shown in Chapter 6.

The differences in impact of the RTT model used in traffic generation, while significant, become negligible when compared to the dramatic differences in impact of the connection structure models used in the experiment.

We found that the application workload model or TCP connection structure has an even more significant effect on all performance metrics than the RTT model used in traffic generation. The two block structure models, representing TCP connections by their sizes alone, create significantly different outcomes for all performance metrics as compared with the *a-b* model that includes object size representation and synchronizations or the *a-t-b-t* model that includes object sizes, the synchronization of objects, and endpoint latencies in its structure. As expected, we found that connection durations and response times increased when epoch structure and endpoint latencies were included in the connection structure model for traffic generation. Also, network-centric measures like the number of active connections in the network increased dramatically as a result of the increase in connection duration.

Unlike RTT models which affected connection duration and response times only up to
1 second, the connection structure models affect these metrics significantly in the body as
well as the tail of the distribution for these metrics.

That is, the distributions for these metrics show significant differences for different connection structure models not only for short connections, but also for very
long connections lasting the entire duration of the experiment. Why is this? We
attribute this effect directly to the fact that connection durations, while affected by
connection RTTs, are most heavily affected by the endpoint latencies, when they
exist, within the connections. Number of active connections in the network also
goes up dramatically, by orders of magnitude, when using endpoint latencies in the
connection structures, as in the case of the *a-t-b-t* model.

In the constrained mode, the absence of endpoint latencies in the block structures and the
a-b model resulted in much heavier queues at the router, thus creating counter-intuitively
long durations and response times because of the second order effects of queuing delay on
connection duration and response times.

In the *unconstrained* mode, for example, using the block structures, the average
connection duration was much smaller than when using the *a-t-b-t* model. This is
because the endpoint latencies in the *a-t-b-t* model add to the duration of connections. However, in *constrained* mode, the long queuing delays caused by the block
structures added long delays to the connection duration; so much so that the duration of connections was longer in some cases for the block structures than for the
a-t-b-t model.

The take away message of this research, if there is to be just one, is that the time components of traffic generation are as important, perhaps more so, than the size components.

While it is important to emulate TCP connections by the size of the connections,
it is equally important to emulate them by their time components. These consist of
the connection RTTs, the sequential or concurrent nature of data exchanges within
connections, and especially the endpoint latencies measured for these connections.

For the bulk of connections in any experiment, window size assignment made no difference
in connection durations or response times.

For a small set of experiments using the control combination of the *a-t-b-t*
connection structure and *usernet* RTT models, we assigned a fixed window size of
8 KB, 16 KB, and 64 KB for all connections in an experiment. This result stated
above is mainly because the bulk of connections are small in size and are unable to
take advantage of the larger windows. For connections carrying more than 1 MB of
data, however, we observed clear differences in these metrics due to different window sizes. These connections performed better with larger window sizes. While this
is to be expected, it is noteworthy that in most Internet traffic, a small number of
connections is found to carry a relatively large percentage of the bytes. Hence for
realistic traffic generation, if those connections had larger window sizes in the original traffic, it is useful to assign them those larger windows during experimentation.
Otherwise, these large connections may not complete in the experiment.

1.10 Organization of Book Chapters

The rest of this book is organized as follows. Chapter 2 presents related works with some background and historical overview of traffic generation and empirical evaluations in networking research. We discuss the three leading traffic generation systems used in empirical research today. We also present some evidence in the literature that point to the need for studying the effects of traffic generation models and path emulations for experimental methodology in empirical networking research. Chapter 3 discusses the design of the various TCP connection structure models in traffic generation used in this study as well as the motivation for their selection. We also present the details of all the RTT methods used in the study and cite their usage in published research wherever applicable. We present the detailed characteristics of the UNC and IBM traffic used in this study.

Chapter 4 presents the details of the network configuration and experimental methodology used in this dissertation for running experiments. This chapter gives details of network setup and how the experiments were designed and conducted. This chapter also presents results for the control combination of connection structure model (the *a-t-b-t* model) and round trip time emulation (the *usernet* method). Chapter 5 presents the main set of results for this study. We present results for connection duration, response times, router queuing, and active connections in the network. Note that the same set of results is presented twice in Chapter 5 for clarity of discussion. First we study the impact of the RTT emulation model; then we study the impact of the application workload model. We discuss results using the UNC and IBM traffic in each of the two network environments – *unconstrained* and *constrained* modes.

Chapter 6 presents additional results. In this chapter, we present results for other connection structures and RTT models that we developed and emulated in our experiments. This chapter also presents results for varying the receiver maximum window size of TCP connections, and discusses the effect on the process of packet arrivals at the router for different connection structure models.

In Chapter 7, we discuss conclusions and future work.

References

1. Andrew L, Marcondes C, Floyd S, Dunn L, Guillier R, Gang W, Eggert L, Ha S, Rhee I (2008) Towards a common TCP evaluation suite. Proceedings of PFLDnet.
2. Hernandez-Campos F (2006) Generation and Validation of Empirically-Derived TCP Application Workloads. Dissertation, University of North Carolina at Chapel Hill.
3. NRC: National Research Council (U.S.) Committee on Academic Careers for Experimental Computer Scientists (1994) Academic careers for experimental computer scientists and engineers. National Academy Press, Washington, DC
4. NS: The Network Simulator. http://nsnam.isi.edu/nsnam/index.php/Main_Page Accessed 15 December 2010.

5. Sommers J, Barford P (2004) Self-configuring network traffic generation. Proceedings of The Internet Measurement Conference
6. Vishwanath KV, Vahdat A (2009) Swing: Realistic and responsive network traffic generation. IEEE/ACM Transactions on Networking
7. Weigle MC, Adurthi P, Hernandez-Campos F, Jeffay K, Smith FD (2006) A tool for generating realistic TCP application workloads in ns-2. ACM Computer Communication Review, 36(3):67–76

Chapter 2
Background and Related Work

A science is any discipline in which the fool of this generation
can go beyond the point reached by the genius of the last
generation.

Max Gluckman
South-African born British social anthropologist (1911–1975)

Experimental networking has evolved significantly over the last two decades, but it remains a daunting endeavor. Throughout this time, traffic generation, a key component for experimental networking, has remained a major challenge. What is traffic generation and what role does it play in empirical networking research? Consider this example: you develop a new Active Queue Management (AQM) scheme for routers on the Internet. AQM is a router-based form of congestion control wherein routers notify end-systems of incipient congestion. The common goal of all AQM designs is to keep the average queue size in routers small [17]. Before deploying this scheme in the wild (Internet), you must test it to ensure that it is better than the existing queue management schemes on your routers. You do this by running experiments using a laboratory network or a simulator.

To produce reliable results from your experiments, you must generate realistic network traffic in your experiments. Why? Say, you use only long-lived FTP-like connections to test your new protocol. While that is representative of some real connections on the Internet, it is not representative of the mix of Internet traffic that will be managed by the router using your new protocol in a production network. Hence, the traffic you generate in the lab or simulator must represent a real mix of traffic on the Internet. So, how do you generate such realistic network traffic? The state of the art in generating realistic traffic today consists of measuring traffic on a real production link and using one of several methods to replay this traffic in the laboratory network. In this study, we use the Tmix traffic generation system to generate traffic in all our experiments. We discuss Tmix and other related work in this chapter.

This chapter is organized as follows: in Section 2.1, we present a brief overview of the network simulators and emulation facilities used by various networking

J. Aikat et al., *The Effects of Traffic Structure on Application and Network Performance*,
DOI 10.1007/978-1-4614-1848-1_2, © Springer Science+Business Media New York 2013

research groups. This is followed by a discussion of the evolution of realistic traffic generation in Section 2.2. In Section 2.3, we present three major traffic generation systems: Harpoon, Tmix, and Swing. In Section 2.4, we present examples in the research that addresses the need to generate realistic background traffic in networking experiments. In Section 2.5, we discuss some community efforts to promote benchmarking tools for congestion control experiments, concluding with a Chapter summary in Section 2.6.

2.1 Network Simulators and Emulation Facilities

Traffic generators are used in network simulators and emulators. Broadly classified, networking experimentation is conducted in two experimental environments: simulation and emulation. Emulation can be further classified into (i) controlled and repeatable experiments in a laboratory, and (ii) live-Internet experimentation. In this section, we shall discuss examples of each of these environments.

At first, the networking research community developed simulators targeted towards the very narrow and specific goals of their projects. Then, from the strong belief that "a diverse set of researchers using a standard framework increases the reliability and acceptance of simulation results" [4] the effort to create the NS network simulator was born almost a decade ago. More recently, several emulation testbed labs have been developed. These include the Emulab [7], Wan-in-Lab [33], ModelNet [19], and UNC's NetLab [27] testbeds.

The most commonly used network simulator is the ns-2 [21] simulator, and ns-3 which is its recently developed replacement. ns-3 is a discrete-event software simulator; that is, the simulation state changes only at discrete points in time. It is a network simulator targeted primarily for research and educational use. It is written in C++ and Python. It is easy to configure and provides an environment for rapid prototyping and building. We use network simulators like ns-2 and ns-3 because they provide complete control, repeatability, and ease of use. However, in doing so, we also sacrifice many protocol implementation details and the realism that requires using real hosts and network elements. Hence, let us discuss some leading emulation testbeds.

Emulab [7], at the University of Utah, is a network testbed, giving researchers a wide range of environments in which to develop, debug, and evaluate their systems. A slice of this lab facility is shown in Fig. 2.1. Emulab is a networked PC cluster that provides a space- and time-shared public facility for studying networked and distributed systems. Emulab tries to transparently integrate a variety of different experimental environments. Historically, Emulab has supported three such environments: emulation, simulation, and live-Internet experimentation. More recently, they have expanded to a fourth environment, virtualized emulation. Emulab allows for *integrated experiments* where they spatially combine real elements with simulated elements to model different portions of a network topology in the same experimental run. This enables new validation techniques and larger experiments than those currently possible using real elements alone [11].

Fig. 2.1 The Emulab Testbed (http://www.emulab.net/, 2010)

Fig. 2.2 The WAN-in-Lab Testbed (http://wil.cs.caltech.edu/, 2010)

WAN-in-Lab [33], at The California Institute of Technology, is an experimental networking testbed aimed at developing, testing and evaluating new communications protocols and technologies. A slice of this lab facility is shown in Fig. 2.2. WAN-in-Lab has a 1500- mile long-haul fiber optic test bed, located in a single laboratory, to allow detailed control and measurement. Initially built to aid FAST TCP research [30], WAN-in-Lab is now used for a variety of networking research and is being equipped to provide a publicly available TCP benchmarking facility. WAN-in-Lab includes a dynamically reconfigurable array of Cisco routers interconnected via OC-48, Gigabit Ethernet (GbE) and 10 Gigabit Ethernet (10GbE) links, using an optical switch. They provide a complement to existing testbeds (that use software for emulating delays) by providing real propagation delay using spools of fiber and active real-time monitoring. Their goal was to reproduce a real production environment more closely.

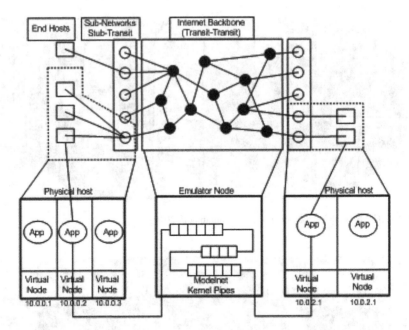

Fig. 2.3 Modelnet in a Testbed (http://www.ics.uci.edu/~mayur/model-net-details.html)

ModelNet [19] at the University of California at San Diego, is a large-scale network emulator that allows users to evaluate distributed networked systems in realistic Internet-like environments. It is a software that can be used as part of a laboratory testbed as shown in Fig. 2.3. With hundreds of applications deployed over the nodes, ModelNet enables them to behave as if they were distributed all over the world. That is, it emulates actual packet delays, losses, and throughput of packets flowing between the different instances of the application. There are physical *Emulator nodes* that run ModelNet on FreeBSD machines, and virtual nodes running applications on Linux machines as shown in the Fig. 2.3. ModelNet also sets up routing tables on the emulator nodes so that packets from two virtual nodes that are on the same physical machine flow through the emulator thus enabling the emulation of a wide-area network testbed.

So, far, we discussed some examples of simulation and emulation environments that provide a controlled, repeatable, and in some cases realistic, systems framework for understanding, testing and evaluating new and existing protocols and algorithms. The third experimentation environment consists of running experiments *in the wild*; that is, running experiments on hosts that are not isolated from the Internet, thus injecting experimentally produced traffic onto real production network traffic. Planetlab is one such overlay testbed that provides real Internet connectivity, and hence does not have the control and repeatability of isolated laboratory testbeds.

PlanetLab is a global research network that began in 2003. Researchers across the globe have used PlanetLab to develop new technologies for distributed storage, network mapping, peer-to-peer systems, distributed hash tables, and query processing.

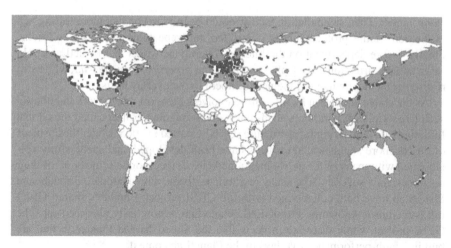

Fig. 2.4 PlanetLab nodes across the globe (http://www.planet-lab.org/, 2007)

PlanetLab currently consists of 1,125 nodes at 511 sites as shown in Fig. 2.4 [24]. It is built as a consortium of academic, industrial, and government institutions. Most of the PlanetLab machines are hosted by research institutions, although some are located in co-location and routing centers (for example, on Internet2's Abilene backbone). All of the machines are connected to the Internet.

All PlanetLab machines run a common software package that includes a Linux-based operating system, mechanisms for bootstrapping nodes and distributing software updates, a collection of management tools that monitor node health, audit system activity, and control system parameters, and a facility for managing user accounts and distributing keys. The key objective of the software is to support distributed virtualization—the ability to allocate a *slice* of PlanetLab's network-wide hardware resources to an application. This allows an application to run across all (or some) of the machines distributed over the globe, where at any given time, multiple applications may be running in different slices of PlanetLab. One of PlanetLab's main purposes is to serve as a testbed for overlay networks. Research groups are able to request a PlanetLab slice in which they can experiment with a variety of planetary-scale services, including file sharing and network-embedded storage, content distribution networks, routing and multicast overlays, QoS overlays, scalable object location, scalable event propagation, anomaly detection mechanisms, and network measurement tools. There are currently over 600 active research projects running on PlanetLab [24].

The advantage to researchers in using PlanetLab (or similar testbeds) is that they are able to experiment with new services under real-world conditions, and at large scale. Of course, the disadvantage is that it is difficult to clearly interpret the results. With far too many unknown and uncontrollable variables when running *experiments in the wild*, it is challenging to draw conclusions. Still, such experiments are valuable and serve an important purpose in empirical networking research as follows. A new protocol could be quickly prototyped and tested for viability in a simulation environment. Then an emulation facility could be used to conduct more testing and evaluation of the protocol under controlled and repeatable network conditions. Finally, before deployment on the Internet, overlay networks like PlanetLab could serve as a confirmation testing platform enabling *experiments in the wild*, while still restricting the deployment of the new protocol to the overlay hosts.

As the above emulation facilities have evolved, the most recent work in building such large-scale networking testbeds has been an ongoing project called the Global Initiative for Networking Infrastructure (GENI), started in 2005. Under the auspices of GENI, more sophisticated testbeds have been developed, and successfully collaborated with many of the above mentioned labs to incorporate some or all of their resources into several large-scale research testbeds. For example, Emulab and PlanetLab have both collaborated with the GENI project. Emulab's shared GENI infrastructure is known as ProtoGENI. PlanetLab is now fully absorbed into the GENI project, while there are projects like the SuperCharged PlanetLab that are building high-performance overlays in the PlanetLab context.

2.2 Evolution of Realistic Traffic Generation

Each of the above mentioned testing and evaluation environments has different properties and goals. However, a common challenge shared among all these environments is the generation of synthetic traffic and the emulation of network path characteristics in experimentation. Floyd and Paxson [9] outlined this problem in the course of declaring traffic generation to be one of the key challenges in modeling and simulating the Internet. Their goal in discussing the difficulties of simulating the Internet was to spur further work in these areas. In a possible response to their challenge, several researchers have attempted to create workload models for traffic generation.

To understand the concerns raised by Floyd and Paxson, consider the simplest method of generating realistic traffic on a single link in the laboratory. One might approximate realistic traffic generation by injecting packets into the network such that the characteristics of these packets are the same as that of the packets on some real link. This is *packet-level traffic generation* and can be achieved in two ways. Either we reproduce the exact sizes and arrival times of every observed packet, or we inject packets into the network such that they preserve some set of statistical properties relevant to the experiment. For example, the packet and byte throughput on the link in 10 millisecond intervals, or the inter-arrival times of these packets could match these same characteristics on some real production link. Such packet-level replay is a straightforward technique that is useful for certain types of experiments. For example, packet-level replays have been used to evaluate cache replacement policies in routing tables [8, 10, 14]. In these experiments, the traffic generated need not respond to the changes in the network. That is, evaluating these policies in the routing tables does not depend on the traffic responding to changes in the policies.

Packet-level traffic generation, however, has two important shortcomings: it is inflexible and it is open-loop. First it is inflexible because there is no way to introduce variability in the experiments. For example, once we acquire a trace, we inject packets into the network to match some characteristics of that trace, as explained above. What if we wished to change packet sizes, or use a different throughput on the link? These are clearly not options available with packet-level traffic replay, other than acquiring a collection of traces and using a different trace (to match the characteristics we need) in different runs of the experiments. Such traffic generation paradigm is simply too cumbersome and impractical for running a large set of experiments [13].

Second, packet-level traffic generation is straightforward. However, since the traffic we replay in our experiments consists of all TCP connections, replaying them in an open-loop manner in the experiments means that we would not preserve the feedback loop that existed between the original sources of the traffic (the endpoints) and the network. TCP is a closed-loop transport protocol. The rate of data transfer is dependent on flow control and congestion control. Flow control is the mechanism used to impose a limit on the maximum sending rate of the sending endpoint. Hence a TCP sender endpoint cannot have more than a maximum, called receiver maximum window, of bytes outstanding (unacknowledged by the receiver endpoint) in the network. Also, the sending rate is limited by a mechanism called congestion control, a set of algorithms at the sender and receiver that react to implicit and explicit feedback from the network. This feedback loop enables the endpoints to react to network congestion. This is important because such reaction itself can change the conditions in the network, thus triggering changes in the behavior of the endpoints. For example TCP traffic reacts to congestion in the network by lowering its sending rate, which in turn decreases congestion. Packet-level replay, however, would not react to changes in the traffic. Therefore, packet-level replay would not be useful in experiments studying the effect of network changes on protocol performance.

Floyd and Paxson strongly urged against open-loop packet-level modeling, and advocated *modeling the sources* of traffic instead [9]; that is, modeling the application behavior at the endpoints. For example, they argued, individual FTP connections between endpoints (sources) must not have a constant rate. Each packet must be sent only after a TCP source receives an acknowledgement for an earlier packet. And if there is congestion in the network, then an FTP connection must vary its sending rate depending on the TCP congestion control window. Also, whether or not there is congestion in the network, different FTP connections will have different average rates, depending on such factors as the TCP window and packet sizes, the connection's roundtrip time, and the congestion encountered in the network. Capturing such application-level interactions and reactions to changing network conditions is essential for realistic traffic generation.

Application workload models are used on top of network stacks which implement flow control and congestion control mechanisms which enable the traffic to react to changes in the network conditions. Such models produce a *closed-loop* traffic generation system which is more realistic. Early application workload models were *infinite source models*. The infinite source model is inherently unidirectional.

That is, for each TCP connection, the sender-receiver pair of generators opens a connection; then the sender constantly sends data packets while the receiver constantly receives or reads these packets. This was a simple model with no parameters and hence was quite popular in leading studies for a number of years, including the mathematical analysis of steady-state TCP throughput [5, 22]. Most long-lasting FTP connections could be represented by this model. This was "realistic" because these FTP connections behaved like real FTP connections on a production link.

The rapid growth of the web drastically changed traffic characteristics on network links so that short (small) request-response exchanges dominated the type of connections on these links. As a result, it was no longer appropriate to use the unidirectional infinite source level model to represent the applications using network links. Such modeling was now unrealistic because most network traffic was found to be bidirectional.

The advent of the web led to attempts by several research groups to model the conversations between web browsers and web servers. One such effort at Boston University led to the development of the SURGE (Scalable URL Reference Generator) model of web traffic [3]. The SURGE model describes the behavior of each user as a sequence of web page downloads and thinktimes between downloads. Each web page download consisted of one or more web objects downloaded from the same server on one TCP connection. Surge models the following components: (i) server file size distribution, (ii) request size distribution, (iii) relative file popularity, (iv) embedded file references, (v) temporal locality of reference, and (vi) idle periods of individual users.

Each component was further modeled by a distribution of values observed for that component. Thus, the empirical distribution for each component was represented analytically. For example, they used the Pareto distribution for modeling the sizes of downloaded objects, and Zipf's law for modeling the popularity of specific pages. Thus, SURGE provided parametric fits for each of the components of the model, heavily relying on powerlaws and other long-tailed distributions.

A model of web traffic contemporary to SURGE was also presented by Mah [18]. It described web traffic using empirical CDFs which were derived from the analysis of packet header traces. They captured traffic on a production link and filtered only HTTP traffic. They modeled the HTTP traffic using parameters of Web client behavior, such as file sizes and think times. They developed empirical probability distributions from those traffic traces to describe various components of the Web client behavior. They then used these distributions to determine a synthetic workload. These components were: HTTP request length, HTTP reply length, document size or number of files per document, think time or time between retrieval of two successive documents, number of consecutive documents retrieved from any given server, and server selection – the parameter used to select each succeeding server accessed. At the lowest level, their model deals with individual HTTP transfers, each of which consists of a request-reply pair of messages, sent over a single TCP connection.

2.3   Current Traffic Generation Systems

Most of the work in workload generation during the 1990s, including the ones we have discussed so far, focused on one or a limited set of application protocols such as FTP, Telnet, and SMTP [23], HTTP [3, 6, 17, 18] and some forms of multimedia. The obvious limitation of these approaches is that real links carry a continuously evolving mix of a number of different applications. While Paxson and Floyd introduced the concept of using source models of individual connections to generate traffic for simulations, they also cautioned that simulating each individual source can be prohibitively expensive in terms of processing time, for many current simulators, because a highly-aggregated Internet link consists (today) of many thousands of simultaneous connections [9].

> Solid, high-level descriptions of aggregate traffic, and simulation models of aggregate traffic that faithfully reproduce the response of the aggregate to individual packet drops (or to other indications of congestion), would be a great help to researchers in exploring large-scale simulations. But, so far, such abstractions are beyond the state of the art. [9]

That was in 2001. Today's state of the art traffic generation systems like Tmix have indeed achieved this goal.

In this section, we discuss three such application workload models used in realistic traffic generation systems. They are the Harpoon model, the Tmix *a-b-t* model, and the Swing model. The Harpoon [25] traffic generator was a landmark contribution in such application workload modeling and traffic generation, because it first addressed the issue of representing a complete set of applications empirically using both TCP and UDP transport protocols without specific knowledge of application protocols or port usage. Swing [29] and Tmix [32] are also empirically based approaches (using *tcpdump* packet header traces) to represent and generate workloads for the complete set of applications using a given network link. Both Swing and Tmix depart from the Harpoon approach by using the additional information available in a packet header trace to represent the internal dynamic structure of connections or flows.

In the rest of this section, we discuss these three leading traffic generation systems.

2.3.1   The Harpoon Model

The Harpoon modeling process was empirically based using easily obtained NetFlow records for all the connections/flows traversing a given network link. Harpoon fundamentally represents a connection or flow by its source-destination IP address pair, its relative start time, and the total number of bytes transferred independently in each direction between source and destination endpoints, as seen at a router. The Harpoon traffic generator [25] takes a router Netflow trace and generates representative packet traffic at the IP flow level.

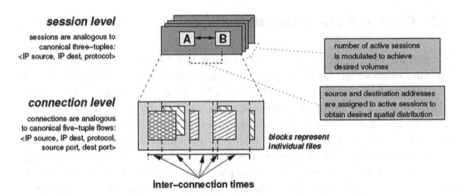

Fig. 2.5 Harpoon's two-level hierarchical traffic model [3]

Sommers et al. define an IP flow as a unidirectional series of IP packets of a given protocol traveling between a source and a destination IP/port pair within a certain period of time. Netflow data includes source and destination AS/IP/port pairs, packet and byte counts, flow start and end times, and protocol information. Harpoon uses this data to generate TCP and UDP packet flows that have the same byte, packet, temporal (diurnal effects associated with traffic volume) and spatial (vis-à-vis IP address space coverage) characteristics as measured at routers in live environments [25].

The Harpoon flow model, as shown in Fig. 2.5 has a two level architecture: connection level and session level. Each "connection" is defined by its file size transferred, and inter-connection time, or time between file transfers. Harpoon connections are 5-tuple flows: source IP address, destination IP address, source port, destination port, and protocol. Harpoon sessions are divided into either TCP or UDP types that conduct data transfers using the respective protocol during the time that they are active. The sessions are 3-tuple flows: source IP address, destination IP address, and protocol.

The session level has two components: the number of active sessions and the IP spatial distribution (IP address space coverage). By modulating the number of sessions that are active at any point in time, Harpoon can match the byte, packet, and flow volumes *every five minutes* from the original data and realize the temporal (diurnal) traffic volumes. Five minutes also happens to be the interval over which flows are aggregated by NetFlow [20]. The intent and domain of Harpoon is to create necessary volumes over longer time scales to produce self-similarity and diurnal patterns in a way that real application traffic is generated.

Thus, the Harpoon model, as summarized in Table 2.1, is made up of a combination of five distributional, empirically-derived, models for TCP sessions: file size, interconnection time, source and destination IP ranges, and number of active sessions. The interval duration parameter was set to five minutes for all their experiments. For UDP packet transfer, Harpoon contains three distributional models: a simple parameterized constant packet rate, a fixed-interval periodic ping-pong, and an exponentially distributed ping-pong. The first source type is similar to some audio and video

Table 2.1 Summary of Harpoon Configuration Parameters for TCP Sources [3]

Parameters	Description
$P_{Filesize}$	Empirical distribution of file sizes transferred.
$P_{InterConnection}$	Empirical distribution of time between consecutive TCP connections initiated by an IP source-destination pair.
$P_{IP\,Rangesrc}$ and $P_{IP\,Rangedest}$	Ranges of IP addresses with preferential weights set to match the empirical frequency distributions from the original data.
$P_{ActiveSessions}$	The distribution of the average number of sessions (IP source-destination pairs) active during consecutive intervals of the measured data. By modulating this distribution, Harpoon can match the temporal byte, packets and flow volumes from the original data.
IntervalDuration	Time granularity over which Harpoon matches average byte, packets and flow volumes.

streams, while the latter two types are intended to mimic the standard Network Time Protocol (NTP) and Domain Name Service (DNS), respectively.

While the Harpoon traffic model was a major breakthrough in empirically derived source modeling, it has its drawbacks. Most importantly, they model the size dimension of application models, completely ignoring the time dimension. As we demonstrate using our results in Chapter 5, the time dimension in application workloads plays a major role in the outcome of experiments. Furthermore, the Harpoon model discards "ACK" flows or flows that are very small, for example, request direction for an HTTP transfer. They also use only complete connections, discarding all incomplete connections, that is, connections for which one or more of the initiation or termination markers (SYN, or FIN/RST) was not recorded in the Netflow logs.

The Harpoon model recreates aggregate trace characteristics without reproducing wide-area network conditions. That is, they do not reproduce connection round trip times, receiver maximum window sizes or loss rates seen on the network. Despite these drawbacks, the Harpoon traffic generator was a landmark contribution because it addressed the issue of representing a complete set of applications using both TCP and UDP transport protocols without specific knowledge of application protocols or port usage.

2.3.2   The Tmix a-b-t Model

Tmix [32], like Harpoon, is also an empirically based approach (using tcpdump packet header traces) to represent and generate workloads for the complete set of applications using a given network link. But Tmix departs from the Harpoon approach by using the additional information available in a packet header trace to represent the internal dynamic structure of connections or flows as follows. Tmix uses inferences about TCP sequence and acknowledgement number exchanges in a packet header trace to characterize connections as sequences of request-response exchanges between endpoints. The request-response exchanges for a connection are

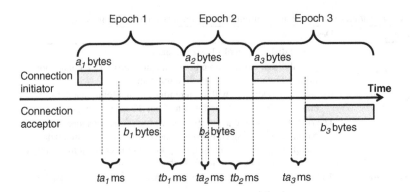

Fig. 2.6 An *a-b-t* diagram illustrating a persistent HTTP connection (sequential)

represented by the number of exchanges, the amount of data in each direction per exchange, and the elapsed time between a request and its response ("server" or intra-epoch latency) or between requests ("user" or inter-epoch latency).

This model allows one to faithfully reproduce the essential pattern of socket reads and writes that the original application performed without knowledge of what the original application actually was. In Hernandez-Campos 2006, the author describes Tmix and demonstrates how the generated traffic displays all the key characteristics of the original captured trace. In addition to the details of request-response exchanges, Tmix reproduces the relative start time, RTT, receiver maximum window size, and loss rate for each connection found in the original tcpdump from a production link.

Thus, starting from a trace of TCP/IP headers collected on a production network, Tmix constructs a model for all the TCP connections observed in the network. The model, a set of *a-b-t* connection vectors, can be used in the workload generator of Tmix to generate the connections and reproduce the application-level behaviors observed on the original network link. The a's and b's are application data units (ADUs) as recorded from the original captured trace, and the t's are the intra-epoch and inter-epoch quiet times within a TCP connection. Modeling as ADUs allows the TCP stack to deal with packetizing, so that inter-packet time is actually not captured, just inter-ADU time is represented.

The *a-b-t* model is used to generate TCP workloads only. A major contribution of this work is that it identifies a fundamental dichotomy in application behavior between connections that exchange data sequentially and those that exchange data concurrently. These two types of connections are shown in Figs. 2.6 and 2.7. Each TCP connection is represented as a connection vector, and every request-response-time sequence is called an epoch within the connection. An epoch represents the abstract characterization of a request/response exchange. Thus every connection consists of one or more epochs.

Unlike Harpoon's model, Tmix's *a-b-t* model is a non-parametric model. Harpoon uses distribution-based models parameterized from analysis of empirical data that are then used with random sampling methods to generate statistically representative

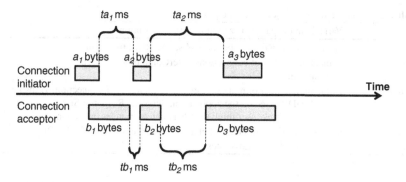

Fig. 2.7 An *a-b-t* diagram illustrating a concurrent connection

workloads in laboratory networks. Tmix, however, emphasizes faithful replays in the laboratory using derived details about each connection to create a replay trace that is used to initiate operations at the socket level to generate workloads. It also offers a method to scale offered loads by sampling the original trace, thus offering huge flexibility in creating modified datasets of workloads resembling the real Internet traffic for specific evaluations on testbeds [13]. This method enables the researcher to introduce controlled load variability in the source-level trace replay experiments without sacrificing realistic workload modeling.

Tmix is a highly flexible traffic generation system and the *a-b-t* model provides a basis for comparing traffic generation methods for our study. Tmix, like the other systems, has its drawbacks. First, it does not emulate UDP flows. Second, Tmix does not account for any correlation among start times of TCP connections; that is, it does not model the sessions that Harpoon and Swing model on top of the connection model.

The Tmix method of traffic generation works as follows. Given a packet header trace, the trace is analyzed and described as a set of connection vectors. Each connection vector describes the application-level behavior of one of the TCP connections in the trace. In addition, each vector includes the relative start time of each connection, and its measured minimum round trip time, the TCP receiver window sizes, and loss rate. The basic approach to generating traffic is to replay each connection vector. For each connection, the replay consists of starting a TCP connection, carefully preserving its relative start time, and reproducing ADUs and inter-ADU quiet times [13].

2.3.3   The Swing Model

Swing [29], like Tmix, is a closed-loop, network-responsive traffic generator that accurately captures the workloads from a range of applications using a simple structural model. But Swing, unlike Tmix or Harpoon, advocates a common parameterization

Table 2.2 Swing's structural model of traffic [29]

Layer	Variable in the parametrization model: Description
Users	ClientIP, numRRE: Number of RREs, interRRE: think time
RRE	numconn: number of connections, interConn: time between start of connections
Connection	Numpairs: number of request-response exchanges per connection, Transport: TCP/UDP based on the application, ServerIP, Response Sizes, Request Sizes, reqthink: user think time between exchanges on a connection
Packet	packetsize, MTU, bitrate, packet arrival distribution (only for UDP)
Network	Link latency, Delay, Loss rates

model for various application classes instead of grouping them all together. Starting from observed traffic at a single point in the network, Swing automatically extracts distributions for user, application, and network behavior. It then generates live traffic corresponding to the underlying models in a network emulation environment running commodity network protocol stacks, generating traces that are statistically similar to the original traces. They extract and assign the following network characteristics: link delays, link capacities, and loss rates.

Swing develops a session model on top of the connection model of Tmix. Swing includes characterizations of the user and session interarrivals which implicitly determine the connection start times. Swing defines request-response exchanges as *RREs*, where a base request for a web page (accompanied by several image downloads as part of that request) and all its responses are considered part of the same *RRE*, but as different connections within the same *RRE*. It could amount to parallel or simultaneous connections. Connections are part of the same RRE if the SYN of a new connection is within an *RREtimeout* of 30 seconds of the previous connection from the same IP address. If not, then this connection is a new *RRE*. However, if this new *RRE* is from the same IP address pair, and if its SYN is within a *session timeout* period of 5 minutes, then it's a new *RRE* in the same session as the previous RRE. If it's beyond the 5 minute session period, then a new session has started.

So, the structural model of Swing, as shown in Table 2.2, is as follows: each session consists of a number of RREs, which in turn consist of a number of protocol connections. Hence their structural model consists of users, sessions, connections, and network characteristics. For each HTTP session, for instance, they pick a randomly generated value (from the corresponding distribution) for each of the variables. First they pick a client and then decide how many RREs to generate along with their interRRE times. For each RRE, they decide how many parallel connections (separated by interConn times) to open and to whom (server). Within a connection they decide the total number of request-response exchanges along with the request sizes, response sizes, and the request think time (reqthink) separating them.

Swing emulates the network path using ModelNet. Every packet is routed to a single ModelNet core. Swing generates traffic that matches the burstiness of the original traffic for both bytes and packets in both directions. They have shown this to be true for a variety of individual applications and original traces at a range of speeds and taken from a variety of locations. The generated traffic also matches burstiness of the packet arrival process of the original trace at a variety of timescales ranging from 1 ms to multiple minutes. Their metrics for success in traffic generation are realism, responsiveness, and maximally random traffic generation. This last metric calls for a traffic generation tool to be able to generate a family of traces *constrained* only by the target characteristics of the original trace and not the particular pattern of communication in that trace. While Tmix strove to generate traffic that was the same as the original traffic, the authors of Swing clearly declare that they want their generated traffic to be "representative" of real traffic and not necessarily the same as the real traffic. Thus Swing was designed to allow experimentation with changing loads and application characteristics. It also allows estimation of experimental variation by generating random instances of traces using different random number seeds.

While Swing is also a highly flexible traffic generation system, it has two major drawbacks. Swing is not application independent like Tmix and Harpoon. Given a packet header trace, they first assign packets and flows to application classes, based on destination port numbers. For those applications with port numbers that cannot be classified, there is an "other" application class. They start with a set of parameters for each application and add in more parameters as needed. This may not be scalable as applications change constantly. However, their argument for doing this is that they can then change the characteristics of the generated traffic in terms of applications represented in the traffic. And like Harpoon, Swing does not use incomplete connections.

In summary, the researchers that developed the Harpoon, Swing, and Tmix workload generators reported extensive validations to show that the resulting synthetic packet-level traffic on an emulated network link was a realistic or faithful reproduction of the traffic seen on a real-world network link. To the best of our knowledge, however, ours is the first research that explores in detail the effects of using different models of application workloads and path characteristics on various metrics of network performance in a realistic network environment.

2.4   Does Traffic Modeling Matter?

Besides the work that has produced realistic application workload modeling and traffic generation tools over the last decade, there have also been a few attempts to show that simply the presence of background traffic (realistic or not) makes a difference in the outcome of the experiments. For example, in [28], the authors show that realistic background traffic matters in experimental evaluations of distributed systems, and that simple models like CBR and Poisson are insufficient. Another example is in [12] where the authors make observations about the effects of background network traffic for TCP protocol evaluations.

In his dissertation [16], Long Le shows that the results for response times using different Active Queue Management (AQM) schemes changes dramatically when a different RTT distribution was used. And in [15], the authors illustrate how variability in network traffic affects buffer dynamics in IP routers. In the rest of this section, we discuss these four research projects more closely.

2.4.1   Does Background Traffic Matter?

In [28], the authors make the point that simple models of background traffic, such as constant bit rate, Poisson arrivals, or deterministic link loss rates are insufficient to capture the effects of background traffic on applications. They contend that we require more complex background traffic models that capture the burstiness on a particular network link. Traffic models that drive tools like Tmix, Harpoon and Swing are based on this idea.

In this paper they show that in order to evaluate distributed systems and networked services in a realistic manner in an experimental testbed, a key ingredient to model correctly is background traffic. They study the impact of background traffic on three applications - web traffic, multimedia traffic, and bandwidth estimation tools. Also, they use four different methods of generating background traffic. They employ constant bit rate (CBR), Poisson model, TCP replay, and Swing. Swing is the only one among these that uses a real trace and generates TCP traffic using stacks on the end-systems. Hence the resulting background traffic using Swing is responsive.

How does this paper relate to the work in this study? They show that realistic traffic matters in experimental evaluations, and that simple models like CBR and Poisson models are insufficient. We move further beyond this idea – we show that even within the realm of realistic traffic models, some aspects of the structural model matter more than others, depending on what is being evaluated. For example, preserving the request-response exchanges within TCP connections affects router queue dynamics, but modeling the inter-epoch times between these request-response exchanges within TCP connections has an even greater effect on router queue dynamics and number of active connections in the network.

2.4.2   Impact of Background Traffic

In [12], the authors examine the effect of background traffic on the performance of existing high-speed TCP variant protocols, namely BIC-TCP, CUBIC, FAST, HSTCP, H-TCP and Scalable TCP. They demonstrate that the stability, link utilization, convergence speed and fairness of the protocols are clearly affected by the variability of flow sizes and round-trip times (RTTs), and the amount of background flows competing with high-speed flows in a bottleneck router.

For all their experiments, they use *dummynet* to assign a per-flow delay. The delay is randomly selected from a distribution obtained from [1]. For background traffic, they use *Iperf* to generate long-lived flows and SURGE to generate short-lived flows. They randomly sample from a distribution of file sizes the amount of data (flow size) to be transferred in each web session. This distribution consists of a log-normal body and a Pareto tail. As an example, for background traffic, they use 12 long-lived flows and SURGE-generated web traffic with 70% body and 30% tail. The minimum file size of the Pareto distribution is 1 MB. The arrival time of flows follows an exponential distribution with intensity 0.6.

Their experimental results include evidence that the presence of some background traffic affects TCP-friendliness. TCP-friendliness is defined to be the fairness of a high-speed flow in sharing bandwidth with another TCP-NewReno or TCP-SACK flow over the same end-to-end path. They do not restrain the maximum window size of TCP-SACK. Their experimental results with no background traffic indicate that with very low RTTs, the TCP-friendliness of H-TCP is the best. All protocols improve their TCP-friendliness at varying degrees when some background traffic is added. Among all the protocols tested, BIC-TCP and STCP show the biggest improvement. There is also significant improvement in the TCP-friendliness of CUBIC under some background traffic.

How does this paper relate to the work in this study? They show that the presence of any background traffic, as opposed to no background traffic, affects certain TCP fairness metrics. While their goal is notable – showing that background traffic matters in protocol evaluation – their traffic is statistically modeled and their methodology for traffic generation lacks the aggregation levels needed to make their case for realistic traffic generation.

2.4.3   Effects of Active Queue Management on TCP Performance

In his dissertation [16], Long Le investigates the effect of active queue management on the performance of TCP applications. This study involves a thorough evaluation of the leading AQM algorithms, including PI, REM, and ARED, comparing them with the prevalent drop-tail queuing in routers. As part of this study, Le, using the same application workload but two different RTT distributions, shows that there are pronounced differences in the response time behavior for almost every AQM scheme using the two RTT distributions. Figure 2.8 shows results using uniform RTT distribution U[10,200] whereas Fig. 2.9 shows results using an empirical RTT distribution [1].

All the distributions for response times, other than the one labeled *uncongested network* are results for experiments with 98% offered load of web traffic. Even the response time CDF (cumulative distribution function) for the uncongested network is quite different for the two different sets of experiments. With uniform RTT distribution, in an uncongested network, 97% of response times are 500 ms or less, whereas with a more general RTT distribution, only 73% of response times are less than 500 ms. When a uniform distribution of RTT was used, there was a pronounced difference in the perceived performance of the different schemes. That is, DCN was clearly the best AQM scheme, followed by PI and REM, then BLUE, ARED and drop-tail in that order. The performances of the last three AQM schemes were significantly worse than the top three.

All else being the same, when the RTT emulation is changed to an empirical, non-uniform distribution, there is virtually no difference among DCN, PI and REM, and though not as good, both ARED and drop-tail are comparable in performance to the other schemes. Also, each of these AQM schemes performed better when using uniform RTT distribution than when using the general distribution.

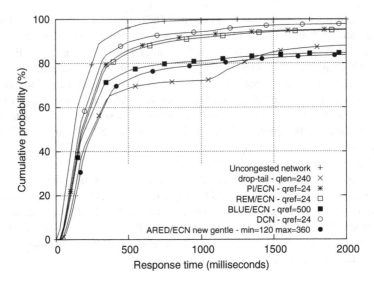

Fig. 2.8 Response Time – uniform RTT Comparison of all AQM algorithms at 98% load [Le05, Fig. 4.112, p. 139]

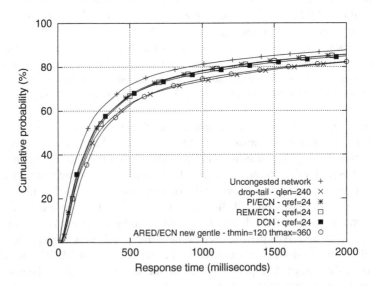

Fig. 2.9 Response Time – empirical RTT Comparison of all AQM algorithms at 98% load [Le05, Fig. 5.54, p. 187]

Although this study about comparing AQM performance also shows that RTT distribution matters for performance evaluations, it does not shed light on what aspects of the RTT model matters, nor does it investigate various RTT models.

How does this study relate to the work in this study? Such studies could strongly encourage or discourage router manufacturers and network administrators from turning on a new queuing algorithm or changing the default TCP congestion control mechanism on the end systems. Hence this only underscores the importance of investigating and developing standards for traffic generation and network emulation. That includes exploring the choice of application workload and network path characteristics in experiments and studying how such choices influence the outcome of these evaluations. Such examples serve to emphasize and underscore our hypothesis that application workload models and network path characteristics greatly influence protocol performance.

2.4.4   TCP/IP Traffic Dynamics and Network Performance

This paper [15] highlights the extent to which assumptions underlying the nature of network traffic can influence practical engineering decisions. Using a simple network configuration of a web server and its clients in the ns2 network simulator, they run experiments to illustrate two points. First, by either implicitly accounting for or explicitly ignoring some aspects of the empirically observed variability of network traffic, a range of different, and at times opposing conclusions can be drawn about the inferred buffer dynamics for IP routers. Second, TCP's feedback-based congestion control is a possible contributing factor to the observed variability of measured TCP/IP traffic over small scales, in the order of RTT.

To show evidence for their first point, they create variability in the workload model as follows. On one end of the spectrum of variability, they use 50 infinite sources that always have data to transfer, thus creating the *no variability* mode. On the other end of the spectrum, they generate purely web workloads similar to SURGE. The main idea behind these Web workload models is that during a Web session, a user typically requests several Web pages, where each Web page may contain several Web objects, thus emulating *high variability* in file sizes. To show evidence for their second point about the TCP feedback loop, they compare the results from simulations using *closed loop* and *open loop* traffic generation.

How does this paper relate to the work in this study? They admit that their network setup and experiments are unrealistic and oversimplified. But through experimental evidence, they emphasize the risk associated with then conventional analysis and simulation of large-scale networks. The risk concerns the wide-spread tendency to rely on and use "a model simplified to the point where key facets of Internet traffic have been lost, in which case the ensuing results are useless (though they may not appear to be so!)." Paxson and Floyd, p. 1043, 1997].

2.5   Community Efforts Toward a Benchmark
for TCP Evaluation

There are currently no standards or benchmarks for protocol evaluation. However, there has been recent interest in the community toward developing better practices for such experimentation. In *Time for a TCP Benchmark Suite?* [31], the authors make one of the first cogent arguments for the need for a TCP benchmarking system. They propose a benchmark consisting of a set of network configurations (topologies and routing matrix), a set of workloads (traffic generation rules), and a set of metrics. The benchmark would have two modes: NS simulation mode, and hardware experiment mode.

More recently, Floyd and Kohler document in their 2008 Internet Draft ("Tools for the Evaluation of Simulation and Testbed Scenarios"), that there has been some effort to formulate evaluation scenarios specific to congestion control experiments. At the same time, there has been increased awareness and consensus among networking researchers for the need to create a common TCP evaluation suite. One of the key components of such a suite would be traffic generation. In Andrew et al. 2008, the authors create a case for a common evaluation standard for TCP evaluations. This paper does not present any results of experimentation, but acts as a powerful catalyst for discussions on this topic. There is also a related and ongoing effort by the "Transport Modeling Research Group" (TMRG) to come up with a consensus for a baseline standard for protocol evaluation. This effort, however, is simply to come up with a consensus, and use that for testing. It does not itself present any experimental results.

While all these efforts are making, albeit small, progress towards benchmarks for TCP evaluations, none of them venture toward the much larger goal of benchmarks for empirical research in networking. This study is a step in that direction.

2.6   Chapter Summary

The above examples (in Sections 2.3, 2.4, and 2.5) are papers or dissertations published within the last few years. There is still no consensus about generating realistic workload models as background traffic in networking research. All of these studies differ from the work in this study significantly, in that we move past the debate of whether or not background traffic matters. Our questions are about the underlying structure of the workload model used in such traffic, and the emulation of path characteristics in such experiments, for network performance. We show, through extensive experimental evidence, how the choices made in both workload modeling and network path characteristics strongly affect network performance for a set of performance metrics.

References

1. Aikat J, Kaur J, Smith FD, Jeffay K (2003) Variability in TCP round-trip times. Proceedings of Internet Measurement Conference, 2003.
2. Andrew L, Marcondes C, Floyd S, Dunn L, Guillier R, Gang W, Eggert L, Ha S, Rhee I (2008) Towards a common TCP evaluation suite. Proceedings of PFLDnet.
3. Barford P, Crovella ME (1998) Generating representative web workloads for network and server performance evaluation. Proceedings of ACM SIGMETRICS.
4. Breslau L, Estrin D, Fall K, Floyd S, Heidemann J, Helmy A, Huang P, McCanne S, Varadhan K, Xu Y, Yu H (2000) Advances in Network Simulation. IEEE Computer, 33(5):59–67.
5. Budhiraja A, Hernandez-Campos F, Kulkarni VG, Smith FD (2004) Stochastic Differential Equation for TCP window size: Analysis and Experimental Validation. Probab. Eng. Inf. Sci., 18(1):111-140.
6. Cao J, Cleveland WS, Gao Y, Jeffay K, Smith FD, Weigle MC (2004) Stochastic Models for Generating Synthetic HTTP Source Traffic. Proceedings of INFOCOM.
7. Emulab: total network testbed. http://www.emulab.net. Accessed 9 July 2011.
8. Feldmeier DC (1998) Improving gateway performance with a routing-table cache. Proceedings of IEEE INFOCOM.
9. Floyd S, Paxson V (2001) Difficulties in simulating the internet. IEEE/ACM Transactions on Networking, 9(4):392–403.
10. Gopalan K, Chiueh TC (2002) Improving route lookup performance using network processor cache. Proceedings of ACM/IEEE Conference on Supercomputing.
11. Guruprasad SB (2005) Issues In Integrated Network Experimentation Using Simulation And Emulation. Dissertation, University of Utah.
12. Ha S, Le L, Rhee I, Xu L (2007) Impact of background traffic on performance of high-speed TCP variant protocols. Computer Networks.
13. Hernandez-Campos F (2006) Generation and Validation of Empirically-Derived TCP Application Workloads. Dissertation, University of North Carolina at Chapel Hill.
14. Jain R (1990) Characteristics of destination address locality in computer networks: a comparison of caching schemes. Computer Networks and ISDN Systems, 18(4):243-254.
15. Joo Y, Ribeiro V, Feldmann A, Gilbert AC, Willinger W (2001) TCP/IP traffic dynamics and network performance: A lesson in workload modeling, flow control, and trace-driven simulations. Proceedings of ACM SIGCOMM.
16. Le L (2005) Investigating the Effects of Active Queue Management on the Performance of TCP Applications. Dissertation, University of North Carolina at Chapel Hill
17. Le L, Aikat J, Jeffay K, Smith FD (2007) The Effects of Active Queue Management and Explicit Congestion Notification on Web Performance. IEEE/ACM Transactions on Networking, 15(6):1217–1230.
18. Mah BA (1997) An Empirical Model of HTTP Network Traffic. Proceedings of IEEE INFOCOM, vol. 2, pp. 592-600.
19. Modelnet: UCSD Computer Science Systems and Networking. http://modelnet.ucsd.edu/. Accessed 4 July 2010.
20. Netflow: Netflow services solutions guide (white paper). http://www.cisco.com/en/US/docs/ios/solutions_docs/netflow/nfwhite.html. Accessed 20 August 2010.
21. NS: The Network Simulator. http://nsnam.isi.edu/nsnam/index.php/Main_Page Accessed 15 December 2010.
22. Padhye J, Firoiu V, Towsley D, Kurose J (2000) Modeling TCP Reno performance: a simple model and its empirical validation. IEEE/ACM Transactions on Networking, 8(2):133–145.
23. Paxson V (1994) Empirically derived analytical models of wide-area TCP connections. IEEE/ACM Transactions on Networking, 2(4):316-336.
24. PlanetLab: The PlanetLab Project. http://www.planet-lab.org/ Accessed 7 July 2010.
25. Sommers J, Barford P (2004) Self-configuring network traffic generation. Proceedings of The Internet Measurement Conference

26. TMRG: The Transport Modeling Research Group. http://trac.tools.ietf.org/group/irtf/trac/wiki/tmrg. Accessed 9 July 2010.
27. UNC-NetLab: University of North Carolina at Chapel Hill's Networking Laboratory. http://www.cs.unc.edu/Research/dirt/ Accessed 10 June 2010.
28. Vishwanath KV, Vahdat A (2008) Evaluating distributed systems: Does background traffic matter? Proceedings of USENIX Annual Technical Conference.
29. Vishwanath KV, Vahdat A (2009) Swing: Realistic and responsive network traffic generation. IEEE/ACM Transactions on Networking
30. Wei DX, Jin C, Low SH, Hegde S (2006) FAST TCP: Motivation, Architecture, Algorithms, Performance. IEEE/ACM Transactions on Networking, 14(6), pp. 1246-1259.
31. Wei DX, Cao P, Low SH (2005). Time for a TCP benchmark suite. https://www.primessf.net/pub/Public/DistributedVirtualTestbed/wei-benchmark.pdf Accessed 10 July 2010.
32. Weigle MC, Adurthi P, Hernandez-Campos F, Jeffay K, Smith FD (2006) A tool for generating realistic TCP application workloads in ns-2. ACM Computer Communication Review, 36(3):67–76
33. WIL: The Wan in Lab project. http://wil.cs.caltech.edu/ Accessed 10 July 2010.

Chapter 3
Workload Modeling and Traffic Generation

> *Building a large packet-switching network is easy;*
> *understanding the behavior of traffic in a large*
> *packet-switching network is nearly impossible.*
>
> Douglas Comer [1]

In this chapter, we present two main topics: traffic characteristics of the input traffic used in this study, and the models developed for this traffic to represent the application workloads and network characteristics. This chapter is organized as follows: in Section 3.1 we give detailed analyses for the traffic characteristics of the two sets of input traffic mixes – UNC and IBM - that we use as input for all our experiments for traffic generation. In Section 3.2 we discuss the Tmix traffic generation system used for all our experiments in this study. In Sections 3.3 and 3.4 we develop the six different connection structure models (application workloads) for TCP connections and the seven different round trip time models (network characteristics) for emulating the end-to-end paths.

3.1 Traffic Characteristics of the Two Input Traces

For realistic traffic generation, we begin with real network traffic captured on production links on the Internet. In this study, we use two very different network traces collected at two diverse locations on the Internet. The first one from UNC was taken on the border link connecting the campus of the University of North Carolina at Chapel Hill to the Internet service provider network. The second trace was taken at an aggregation switch for four internal networks, connecting one of IBM Corporation's largest development sites to the Internet. The UNC campus trace was a 1-hour packet-header trace taken on a weekday during the school year, from 2:00 PM to 3:00 PM on January 10, 2008.

J. Aikat et al., *The Effects of Traffic Structure on Application and Network Performance*, 37
DOI 10.1007/978-1-4614-1848-1_3, © Springer Science+Business Media New York 2013

The IBM trace was also a 1-hour packet-header trace which was representative of typical peak workday traffic on their corporate network, and was taken from 2:20 PM to 3:20 PM on October 10, 2006. Both these traces were captured using a 1Gbps Endace Systems' DAG capture card on a FreeBSD monitoring machine which is a 1.8 GHz server class PC with 1.2 GB of memory. DAG technology provides 100% capture into host memory at full line rate for all packets on the link [2]. The traffic captured by the monitor was then converted to *pcap* and processed using an enhanced *tcpdump* program, and several diagnostic and other tools developed at UNC.

In the rest of this section, we present detailed characteristics for the UNC and IBM traffic. Why? While most network researchers will agree that application workload modeling is essential for realistic traffic generation, we also know that there is no such thing as a *standard* network trace. That is, two large production links on the Internet will likely yield two slightly different, or as in our case two fairly different, traffic mixes. While we use such real traffic as input for our empirical studies, we emphasize by example that we must first analyze and understand the characteristics of the input traffic. Yes, indeed, the characteristics of the input traffic play a major role in experimental outcomes. Using two such input traffic mixes then validates the results more firmly while helping to bring out any methodological choices that lead to differences in the results from using the two different inputs. Hence it is also useful to clearly study the similarities and differences between the two input traces.

3.1.1   Throughput

We begin the analysis of the two input traffic mixes by presenting the time series of their throughput.

In Figs. 3.1 through 3.4, we show the time series of the link throughput in both directions for the original UNC and IBM traces respectively as captured. Although both traces were an hour long, we have shown only the period between 10 and 50 minutes because that is the period we use for all our experiments. The UNC original traffic, as captured, averaged 533 Mbps in one direction (labeled *high*) and 248Mbps in the other direction (labeled *low*). The IBM original traffic, as captured, averaged 464 Mbps in the high throughput direction and 427Mbps in the other direction. Both exhibit variability, but the IBM traffic is significantly more variable.

The throughput in these figures is for TCP traffic only. However, we do not use all of these connections to generate traffic in our experiments. We process this traffic as follows. First, using tools developed by Hernandez-Campos [3], we classify the captured traffic into two categories of TCP connections. The first category of connections, we discard. These connections consist mainly of two sets: one in which the connections and their packets carry no data, and second in which the connections were captured in only one direction. The first set of discarded connections and packets carried no data, and the second carried small amounts of data in only one direction. For connections that carried data in only one direction,

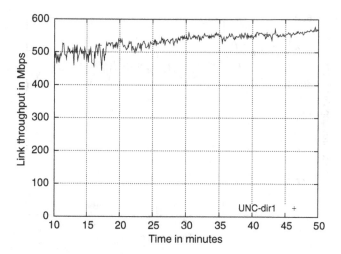

Fig. 3.1 Throughput as captured (high) – UNC

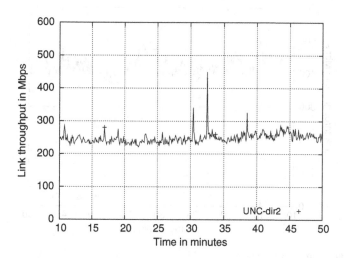

Fig. 3.2 Throughput as captured (low) – UNC

we included them if we captured the packets traversing both directions for that connection. Connections that carried no data, however, still contributed a good fraction of the throughput due to their packet overhead. For example, in the UNC traffic, the connections carrying no data were 10% of the total connections.

The connections with packets seen in only one of the two directions constituted 16% of the total connections though they carried only 1.8% of the total data. And in the IBM traffic, the connections carrying no data were 7% of the total connections. The connections with packets seen in only one of the two directions constituted 5.6%

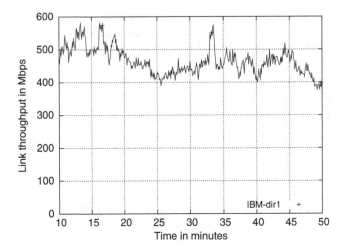

Fig. 3.3 Throughput as captured (high) – IBM

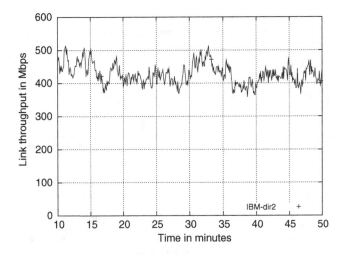

Fig. 3.4 Throughput as captured (low) – IBM

of the total connections and carried negligible (close to 0%) of the total data. It would be interesting to study what applications were represented by these discarded connections, but that is out of scope of this study.

The second category of connections constitutes the traffic we use for emulation in our experiments. This is the bulk of the captured traffic that we then classify into sequential and concurrent connections (see Section 3.1.2 for their representations). We further classify the sequential and concurrent connections into complete and incomplete connections. A complete connection is one for which we see the SYN and FIN or RST for the connection. An incomplete connection is one in which we

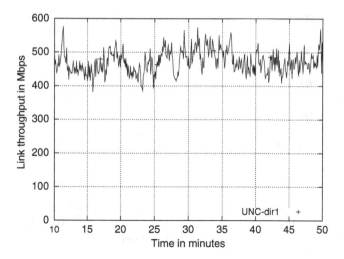

Fig. 3.5   Offered Load (high) – UNC

do not see any one or more of these initiation or termination markers for that connection.

In Hernandez-Campos 2006, the authors used only complete connections for their study. We extend that work of traffic generation by also including incomplete connections since these connections form a large part of the captured traffic. For example, for the UNC trace, 70% (about 4.5 million) of the connections were complete sequential connections, carrying 52% of the total data bytes. And while only 0.37% (about 24,000) of the connections were incomplete concurrent connections, these connections carried fully 21% of the total data bytes. Similarly, for the IBM trace, 80% (about 2.6 million) of the connections were complete sequential connections, carrying 56% of the total data bytes. And while only 0.63% (about 20,000) of the connections were incomplete concurrent connections, these connections carried fully 12% of the total data bytes.

After including all the sequential and concurrent TCP connections, both complete and incomplete, the UNC trace had nearly 4.7 million total connections with an average offered load of 471 Mbps in one direction and 202 Mbps in the other, as shown in Figs. 3.5 and 3.6. The IBM trace had about 2.8 million connections with an offered load of 404 Mbps in one direction and 366 Mbps in the other, as shown in Figs. 3.7 and 3.8.

3.1.2   Sequential and Concurrent Connections

Of the 4.7 million total connections in the UNC traffic, 4,568,847 are sequential connections and 115,045 are concurrent connections. The sequential connections

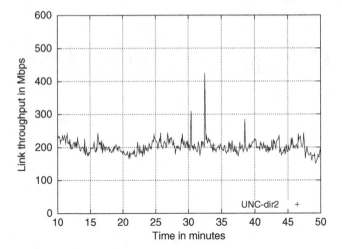

Fig. 3.6 Offered Load (low) – UNC

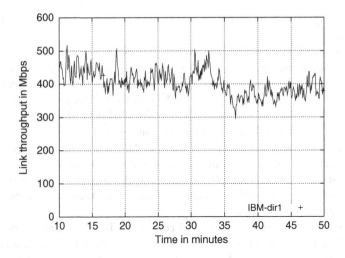

Fig. 3.7 Offered Load (high) – IBM

transfer 214 billion bytes (70%) of the total payload. The concurrent connections transfer 86 billion bytes (28%) of the total payload. Of the 2.8 million total connections in the IBM traffic, 2,733,996 are sequential connections and 51,058 are concurrent connections. The sequential connections in the IBM traffic transfer 310 billion bytes (85%) of the total payload. The concurrent connections transfer 55 billion bytes (15%) of the total payload.

So what are sequential and concurrent connections? Hernandez-Campos et al. first identified and classified TCP connections for traffic generation as being sequential or

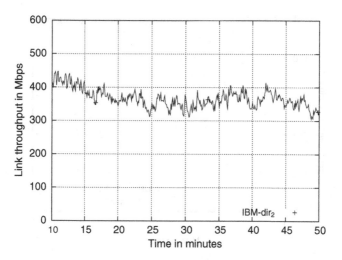

Fig. 3.8 Offered Load (low) – IBM

concurrent in nature. A sequential TCP connection is a sequence of one or more request-response exchanges, called epochs. Each epoch describes the properties of a pair of application data unit (ADU) exchanges between the two TCP endpoints. [3]. The concept of an epoch arises from the client/server structure of many distributed systems, in which one endpoint acts as a client and the other one as a server. This representation captures the sequential order of the ADUs within the TCP connection, the direction in which the ADUs flow, and the sizes of the ADUs.

In the sequential model, the application data is either flowing from the client (connection initiator) to the server (connection acceptor) or from the server to the client. However, some TCP connections are not driven by this client-server model of data exchanges. Some applications send data from both TCP endpoints simultaneously. For example, such connections are said to have *data exchange concurrency* and are called concurrent connections. In such connections, one or more pairs of ADUs are exchanged simultaneously.

3.1.3 Application-level Characteristics

Hernandez-Campos et al. first developed this classification for all TCP connections into sequential and concurrent connections with the goal of capturing and generating application data exchanges, including the pattern of such exchanges, without knowledge of the underlying applications. In this sub-section, we present data for these application-level characteristics for the two input traffic mixes obtained from their packet-header traces. Specifically, we compare the distributions for the number of epochs per connection, the size of ADUs, and the endpoint latencies in the connections for the two traces.

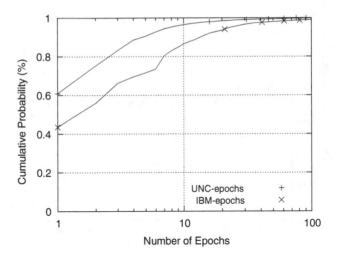

Fig. 3.9 Number of connection epochs UNC and IBM – CDF

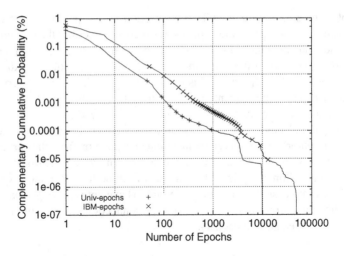

Fig. 3.10 Number of connection epochs UNC and IBM - CCDF

3.1.3.1 Epochs

An epoch is a request-response exchange within a sequential TCP connection. On average, the sequential connections in the UNC trace used 3 epochs to transfer bytes, with a standard deviation of 42 epochs. Sequential connections in the IBM trace used a mean of 9 epochs with a standard deviation of 123 epochs to transfer data. The cumulative distributions of number of epochs per connection for both traces are shown in Figs. 3.9 and 3.10. The CCDFs for both distributions are

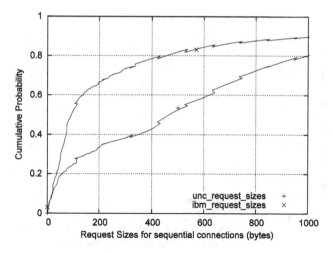

Fig. 3.11 Request sizes - sequential connections

approximately linear on a log-log scale. Hence, both clearly have a heavy-tailed distribution in the number of epochs.

We observe that 60% of sequential connections in the UNC trace had only one epoch, with 90% of sequential connections having only 5 or fewer epochs. For the IBM trace, 44% of sequential connections had only one epoch, with 90% of sequential connections having 14 or fewer epochs. Only 3% of UNC connections had 12 or more epochs whereas 13% of IBM connections did. So, while the top 3% of UNC connections had 12 or more epochs, the top 3% of IBM connections had 33 or more epochs. In the UNC trace, only 0.01% of connections had 1000 epochs or more, whereas in the IBM trace that number was 0.05% of connections. The tails of the distributions clearly show that the number of epochs in connections in the IBM trace was much higher than those in the UNC trace.

3.1.3.2   Application Data Units (ADU)

Sequential connections exchange data in epochs, that is, in a request-response pattern. Hence, we measure the ADU sizes in each epoch as a request-size and a response-size for these sequential connections. Concurrent connections, on the other hand, send bytes in both directions simultaneously, so we represent all concurrent ADUs in one distribution. In Figs. 3.11 through 3.16, we show the cumulative distributions of these ADU sizes for sequential and concurrent connections. First, let us compare the request sizes for both the UNC and IBM traces in Figs. 3.11 and 3.12.

The median data size for requests in sequential connections is 460 bytes in the UNC trace, but only 84 bytes in the IBM trace. 20% of these requests are greater than 1000 bytes in the UNC trace but only greater than 466 bytes in the IBM trace.

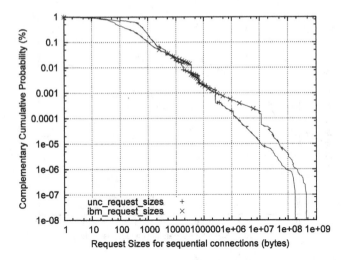

Fig. 3.12 Request sizes - sequential connections

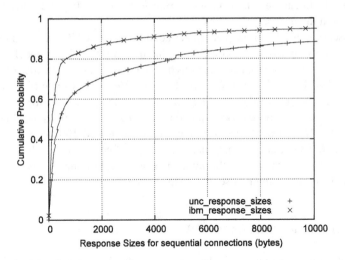

Fig. 3.13 Response sizes - sequential connections

But the average request size is 2.5 KB in the UNC trace and 6 KB in the IBM trace. The top 10% of request sizes are greater than 1.6 KB in the UNC trace and greater than 1 KB in the IBM trace. So, while the IBM traffic has a few sequential connections with very large request sizes (skewing the average), most of the request sizes in the UNC trace are comparatively larger.

Now, let us compare the response sizes for the two traces in Figs. 3.13 and 3.14. The median response size is 420 bytes in the UNC trace and 128 bytes in the IBM

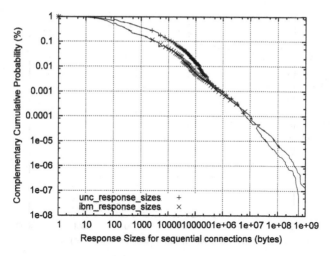

Fig. 3.14 Response sizes - sequential connections

trace. 20% of responses are greater than 4 KB bytes in the UNC trace but only greater than 680 bytes in the IBM trace. The average response size is 11 KB in the UNC trace and 9 KB in the IBM trace. The top 10% of response sizes are greater than 13 KB in the UNC trace but greater than only 3.3 KB in the IBM trace. So we note that sequential connections in the IBM trace have much smaller response sizes as compared with those in the UNC trace. The CCDFs clearly show a heavy-tailed distribution for response sizes in both UNC and IBM traffic.

For concurrent connections, we consider all ADUs in one distribution, since there are no request-response exchanges within these connections. As shown in Figs. 3.15 and 3.16, the median size of concurrent ADUs is 208 bytes in the UNC trace and 91 bytes in the IBM trace. 20% of ADUs are greater than 1400 bytes in the UNC trace but only greater than 610 bytes in the IBM trace. The average ADU size is 5.9 KB in the UNC trace but larger than 11.5 KB in the IBM trace. As with the sequential ADUs, we see here that a small number of very large concurrent ADUs skew the average ADU size in the IBM traffic.

The top 10% of ADU sizes are greater than 6.8 KB in the UNC trace and greater than only 3.4 KB in the IBM trace. ADU sizes in the IBM trace, excluding a few very large ADUs, are smaller than those in the UNC trace. Figure 3.16 shows that the ADU sizes in both sets of concurrent connections are equivalent in the tail, and they have a heavy-tailed distribution.

3.1.3.3   Endpoint Latencies

In the Tmix *a-b-t* model, besides ADUs, the sequential and concurrent connections have endpoint latencies. We identify two kinds of such endpoint latencies, developed

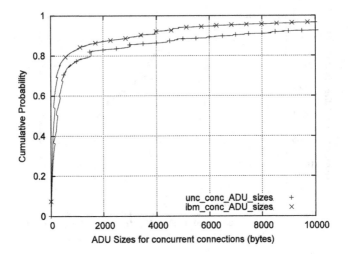

Fig. 3.15 ADU sizes - concurrent connections

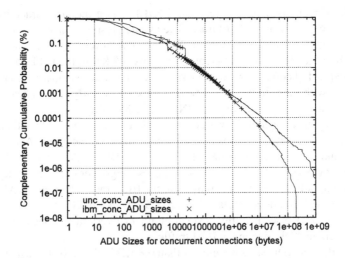

Fig. 3.16 ADU sizes - concurrent connections

as part of the *a-b-t* model by Hernandez-Campos et al. First, we have the intra-epoch endpoint latency which is the time elapsed at the connection initiator (client), and within an epoch, between sending a request and receiving its response. This is usually the time taken by the server to process the request plus one round trip time of network latency. Second, we have the inter-epoch endpoint latencies which are the times between two epochs, that is, the time between receiving a response and sending the next request. These could be due to either some processing time or user

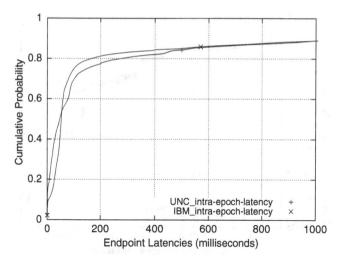

Fig. 3.17   Intra-epoch endpoint latencies for sequential connections

thinktime. Every epoch in a sequential connection has an intra-epoch latency. And multiple epoch connections have inter-epoch latencies as well. Concurrent connections have one or more endpoint latencies. These latencies are simply associated with the preceding ADU sent by that endpoint. When endpoint latencies are less than 500 *ms*, they could easily be due to network effects and hence we do not consider them as part of the source-level behavior. Hence we do not emulate endpoint latencies less than 500 *ms*.

It is worth noting that of all the measured latencies, roughly 16% of intra-epoch latencies were greater than 500 *ms* for both UNC and IBM traces. This means that server processing latencies have an effect on a small number of epochs in both traces. For inter-epoch latencies 44% of them were larger than 500 *ms* for the UNC trace, but only 20% of them were larger than 500 *ms* for the IBM trace. Each latency measure is considered a data point here, regardless of the number of latencies measured for each connection. This difference in inter-epoch latencies between the two traces becomes very significant when we study the effect on queue length. For the same level of capacity constraint on the router-to-router link (95%), the IBM trace shows much heavier queues using this model of connection structure because in the UNC traffic, the larger number of inter-epoch latencies plays a significant role in allowing the queue to drain and maintain a smaller queue overall. For concurrent connections, it is interesting to note that most (99%) of the latencies were greater than 500 *ms* for both UNC and IBM traces. So essentially all measured latencies are emulated for concurrent connections.

Let us now analyze these endpoint latencies in the UNC and IBM traffic. We show all the intra-epoch latencies (including those below 500 *ms*) for the two traces in Figs. 3.17 and 3.18. Note that for the CCDFs, we start the Y-axes at 500 *ms*. Each plot compares the data for the two traces – UNC and IBM.

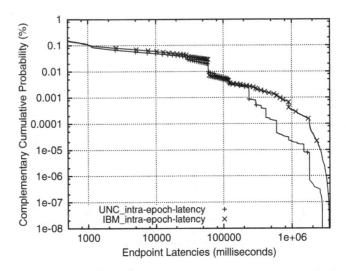

Fig. 3.18 Intra-epoch endpoint latencies for sequential connections

We observe that the median intra-epoch latency is 47 milliseconds for UNC connections and 53 milliseconds for IBM connections. 20% of these latencies are greater than 273 milliseconds for UNC connections and greater than 168 milliseconds for IBM connections. The average intra-epoch latency is quite high, however, with 3.1 seconds for UNC and 4.7 seconds for IBM connections. This high average reflects the small number of multi-epoch, long connections with long latencies present in both traces. The top 10% of intra-epoch latencies are greater than 1 second for the UNC trace and greater than 1.1 seconds for the IBM trace.

We now compare all inter-epoch latencies for the two sets of traffic in Figs. 3.19 and 3.20. We observe that the median inter-epoch latency is 173 milliseconds for UNC connections and 55 milliseconds for IBM connections. 20% of the inter-epoch latencies are greater than 1.5 seconds for UNC connections and greater than 490 milliseconds for IBM connections, thus much longer than their respective intra-epoch latencies. The average latency is also quite high with 5.6 seconds for UNC and 5.9 seconds for IBM connections, again reflecting the small number of multi-epoch, long connections with long endpoint latencies present in both traces. The top 10% of inter-epoch latencies are greater than 7.5 seconds for the UNC trace and greater than 3 seconds for the IBM trace.

Finally, let us compare all the endpoint latencies for concurrent connections for the two traces in Figs. 3.21 and 3.22. Here, we observe that the median latency for concurrent connections is 1.1 seconds for UNC connections and 1.5 seconds for IBM connections. 20% of these latencies are greater than 4 seconds for UNC connections and greater than 17 seconds for IBM connections. Note that concurrent connections constitute only a small fraction of the total number of connections in both traces – 1.8% of UNC connections and 1.6% of IBM connections, but they transfer 28% and 15% of the total bytes respectively. These percentages stated here

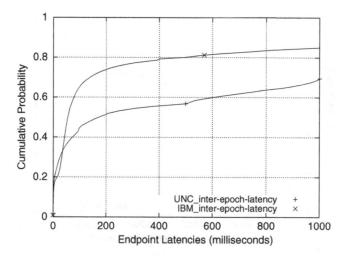

Fig. 3.19 Inter-epoch endpoint latencies for sequential connections

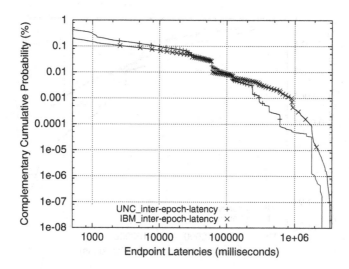

Fig. 3.20 Inter-epoch endpoint latencies for sequential connections

for endpoint latencies are for connections within that fraction, and not part of all the endpoint latencies. Still, these long latency concurrent connections clearly carry a large number of bytes. In the case of the IBM trace, they contribute to the heavier distribution of connection durations, compared with that of UNC connections. The average latency is also quite high, with 6.7 seconds for UNC and 16.7 seconds for IBM connections. The top 10% of latencies in concurrent connections are greater than 14 seconds for the UNC trace and greater than 60 seconds for the IBM trace.

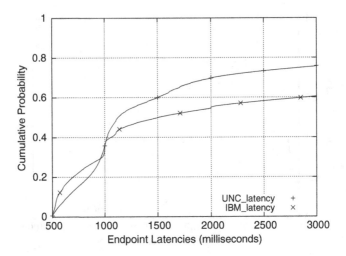

Fig. 3.21 Endpoint latencies for concurrent connections

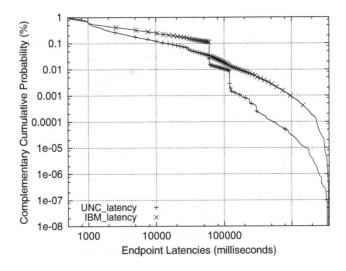

Fig. 3.22 Endpoint latencies for concurrent connections

We must note here that for both sequential and concurrent connections in the original trace, there are connections which exhibit a quiet time between the last ADU and TCP's connection termination. Most of these quiet times are under 500 *ms* and hence discarded anyway. However, there are a few connections with exceedingly long quiet times at the end. Such quiet times reflect more realistic durations for those connections, but add much overhead to our traffic generation system. Hence, we do not model any quiet times that occur after the last ADU within a connection.

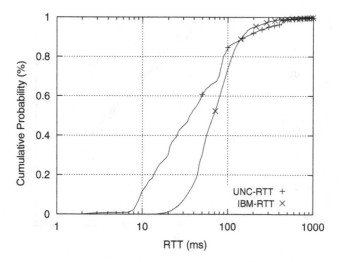

Fig. 3.23   CDF of round trip times

3.1.4   Network-level Characteristics

So far, we discussed the application-level characteristics for the UNC and IBM traffic that we use as input for generating traffic in all our experiments. We now discuss the network-level characteristics of round trip times and window sizes for these TCP connections. For this discussion, we do not distinguish between sequential and concurrent connections, but treat all connections as simply TCP connections.

3.1.4.1   Round Trip Times (RTTs)

The *round-trip time* (RTT) of a TCP connection between two endpoints, a sender and a receiver, is defined as the time it takes for a TCP segment from the sender to reach the receiver and for a segment carrying the generated acknowledgment from the receiver to return to the sender.

The cumulative distribution functions (CDFs) for the measured minimum round trip time per connection in the two traces are shown in Fig. 3.23. The CCDFs for the same are shown in Fig. 3.24. The RTTs in the UNC trace were on average smaller than those in the IBM trace, but the CCDF shows much longer connection RTTs for the UNC trace than in the IBM trace in the tail of the distributions. The mean RTT for connections was $80\,ms$ in the UNC trace while it was $92\,ms$ in the IBM trace. The standard deviation of RTTs was $210\,ms$ and $144\,ms$ for connections in the UNC and IBM traces respectively.

Thus while the median RTT for connections in the UNC trace was $36\,ms$, fully 3% of these connection RTTs were above $429\,ms$. And while the median RTT for connections in the IBM trace was $68\,ms$, the top 3% of these connection RTTs were

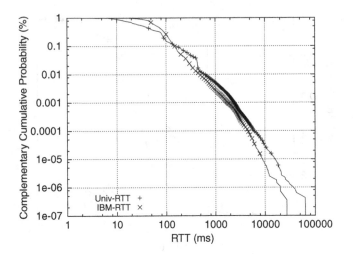

Fig. 3.24 CCDF of round trip times

greater than 275 *ms*. As seen in Fig. 3.24, some connection RTTs were longer than one second for both UNC and IBM connections. Such long delays are sometimes due to compounding effects of long propagation delays added to slow modems on one end of the connection, or due to long delays on cell hosts in the network.

We used these empirical measures to develop all our RTT models discussed later in this chapter.

3.1.4.2   Receiver Window Sizes

Just as we used the empirical measures from the original RTT distributions to develop our RTT models, we used the empirical measures from the original receiver-advertised maximum window size distributions to develop the window size models for our experiments. Hence, let us now examine this network-level characteristic in the two input traces. But first, what is the role of the receiver window size in a TCP connection? When a segment is received by a TCP endpoint, its payload is stored in an operating system buffer until the application uses a system call to receive the data. In order to avoid overflowing this buffer, TCP receiver endpoints use a mechanism called *flow control* to impose a limit on the maximum sending rate of the sending endpoint. Hence a TCP sender cannot have more than this maximum, called receiver maximum window, of bytes outstanding (unacknowledged) in the network.

How is this relevant to our traffic generation? Window size allocation in TCP connections affects the growth of the window of unacknowledged packets that the sender can have in the network. Hence a larger receiver window size, all other thing being equal, means that a TCP connection can transmit data faster and have more data in the network before it receives feedback from the other end of the connection.

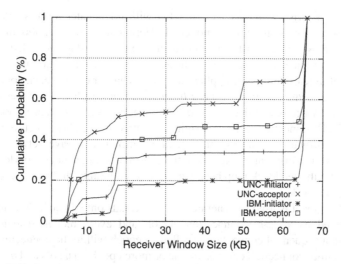

Fig. 3.25 CDF of receiver maximum window sizes of the input UNC and IBM traces

In this study, we measured the maximum advertised window for both ends in each connection from the original trace for both UNC and IBM traffic. Each connection in all our experiments, unless otherwise specified, was assigned the measured receiver window for each of the two endpoints of the TCP connection. This included all the experiments regardless of the connection structure model used for traffic generation, the RTT model used for network emulation, or the network environment, that is *unconstrained* or *constrained* link mode, for each experiment. This maximum receiver window is often different for each endpoint of a connection. Hence we show the separate distributions for the initiator of a TCP connection, and the acceptor for that connection. We show this data for both the UNC and IBM traces in Fig. 3.25.

We observe from this figure that the smallest maximum receiver window size is 4 KB for any connection in both traces. This window size then increases to values of 8 KB, 16 KB, 32 KB, 48 KB, or 64 KB. We did not measure or analyze window scaling and hence the maximum window size we measured was 64 KB. As shown in this figure, the initiators have larger advertised window sizes – roughly 65% of UNC initiators had 64 KB and 80% of IBM initiators had 64 KB receiver windows. However, only 25% of UNC acceptors and 50% of IBM acceptors had 64 KB receiver windows.

3.2   Traffic Generation with Tmix

For all experiments in this study, we use the Tmix traffic generation system. Although we discussed some details about this system among the related works in Chapter 2, let us briefly discuss the Tmix model for both application workload and network characteristics in this section. This will aid in later discussions in this chapter when we present our own models. In Hernandez-Campos 2006, the author developed a new application-level model to characterize workloads, called the *a-b-t* model. Given a packet header trace collected from an arbitrary Internet link, this work algorithmically infers the application-level behavior driving each connection, and casts it into the notation of the *a-b-t* model. The result from processing the packet header trace is a collection of *a-b-t* connection vectors, each vector representing a TCP connection from the original captured trace. These vectors are then replayed in software simulators and testbed experiments to drive network stacks. This replay of the original traffic, using the *a-b-t* model, generates workloads that fully preserve the feedback loop between the TCP endpoints, and also preserve the state of the network.

The *a-b-t* model is used to generate TCP workloads only. Each TCP connection is represented as an *a-b-t* connection vector, and every request-response-time sequence in a sequential connection is called an epoch within the connection. Thus every sequential connection consists of one or more epochs. The *a*'s and *b*'s in both sequential and concurrent connections are the application data units (ADUs), sizes as recorded from the original captured trace. The a-type ADUs are data units sent from the connection initiator to the connection acceptor, and the b-type ADUs are data units sent from the connection acceptor to the connection initiator, i.e. data flow in the opposite direction. The *t*'s represent the quiet times during which no data segments are exchanged. The quiet times may be the time taken between sending of ADUs to the transport layer, or it may be user think times or server processing times. The reason for the quiet times and the actual data in the ADUs are not important to traffic generation, but modeling these quiet times and application data exchanges is very important to represent the lifetime of the connection as we will see in this study. This *a-b-t* emulation model faithfully reproduces the essential pattern of socket reads and writes that the original application performed without knowledge of what the original application actually was. Furthermore, Tmix emulates network path characteristics by assigning to each connection its observed minimum RTT and receiver window sizes.

We note here that there is a fundamental difference between Tmix and the other two traffic generation systems [4, 6] discussed in Chapter 2. Although all three systems are based on modeling traffic and network characteristics from empirical measures of real network traffic, Tmix is a non-parametric model of traffic generation. Tmix accurately and faithfully replays the application-level behavior using a set of connection vectors using real TCP sockets on the traffic generators. Each connection vector input to the traffic generators represents exactly one TCP connection from the original traffic and there is a one-to-one assignment of connection parameters for each connection from the original traffic to the replayed traffic. On the other hand, both Harpoon and Swing use parametric modeling; they are based on random sampling from distributions of empirical parameters of network traffic.

3.3 Variations in the Workload Model

The Tmix *a-b-t* model is a complete representation of a connection's structure for traffic generation. We define connection structure as the representation of the connection workload that has one or more of the following components: ADU sizes, connection sizes, direction and sequence of ADUs, and endpoint latencies. Henceforth, we call this full Tmix model as the *a-t-b-t* model to show its complete representation. Our *a-t-b-t* model is the same as Tmix's original *a-b-t* model. As described in Chapter 2, the *Harpoon* traffic generation system uses a very different and much simpler model for modeling TCP connections. While the Tmix model includes every application data unit and quiet time within a connection, the *Harpoon* model simply represents each connection as two blocks of data transferred, one in each direction. Hence we begin with this simple model for representing a TCP connection.

To study the effect of different connection structures on application and network performance, we developed six different structural models to represent a TCP connection. Our six models, as discussed below, were developed with a representation of the simple *Harpoon* model on one end of the spectrum and the *a-t-b-t* model on the other end. We now present all six connection structure models as originally developed in this study. Although only four of these models were used for the complete set of experiments presented in Chapter 5, we ran smaller subsets of experiments with all of them.

Since the same traces of the original traffic were used to build each of these models, we define some notations for clarity and consistency using the *a-b-t* model representation shown in Figs. 3.26 and 3.27 for sequential and concurrent connections respectively. Let the sequential connection shown in Fig. 3.26 with three epochs be represented by the following connection vector $\{ (a_1, ta_1, b_1, tb_1), (a_2, ta_2, b_2, tb_2), (a_3, ta_3, b_3, tb_3)\}$.

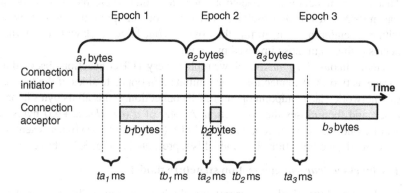

Fig. 3.26 An *a-b-t* diagram illustrating a persistent HTTP connection (sequential)

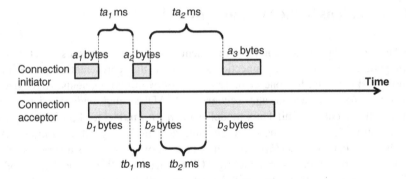

Fig. 3.27 An *a-b-t* diagram illustrating a concurrent connection

For the concurrent connection shown in Fig. 3.27, let the connection vector be represented as (α, β) where $\alpha = \{ (a_1, ta_1), (a_2, ta_2), (a_3, ta_3) \}$ and $\beta = \{ (b_1, tb_1), (b_2, tb_2), (b_3, tb_3) \}$. In both cases, let $a = a1 + a2 + a3$ be the total bytes transferred by the original connection initiator to the connection acceptor. And let $b = b1 + b2 + b3$ be the total bytes transferred by the original connection acceptor to the connection initiator.

We use these notations to describe the six connection structure models below. To use the Tmix traffic generation system for running experiments using these new models, we made some changes as follows. We modified the input connection vectors to the Tmix system to include accurate representations for each of our new models, and we modified the replay engine to appropriately parse the new models and replay the TCP connections in our experiments.

(i)   The *Harpoon* connection structure model

Harpoon models a TCP connection by its size and direction of data transfer. That is, a connection is modeled as two endpoints where the first endpoint transmits X bytes while simultaneously the second endpoint transmits Y bytes with both endpoints transmitting their bytes as one large block without internal delays (other than those imposed by TCP).

Hence in our *Harpoon* model, we replay every TCP connection observed in a trace as two TCP connections, each initiated on opposite sides of the laboratory network. Each endpoint opens a TCP connection, sends all its bytes in one block and then closes the connection. A total of a $(a = a1 + a2 + a3)$ bytes is sent by one TCP connection, and a total of b $(b = b1 + b2 + b3)$ bytes is sent by the other TCP connection. This model is represented in Fig. 3.28 above.

(ii)   The *block-concurrent* connection structure model

We developed the *block-concurrent* model (shown in Fig. 3.29) as a variation of the *Harpoon* model. Unlike the *Harpoon* model, however, a TCP connection observed in a trace is represented in this model by only one TCP connection between two endpoints. The two blocks, a $(a = a1 + a2 + a3)$ bytes and b

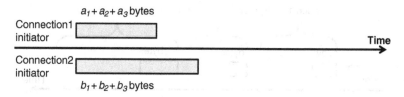

Fig. 3.28 The *Harpoon* connection structure model for all TCP connections

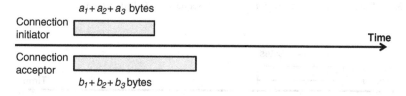

Fig. 3.29 The *block-concurrent* connection structure model for all TCP connections

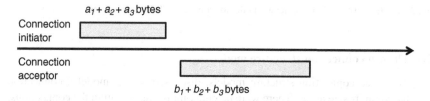

Fig. 3.30 The *block-sequential* connection structure model for all TCP connections

$(b=b1+b2+b3)$ bytes, are sent simultaneously by the two endpoints after connection establishment. In this model, all the TCP connections in the experiment behave like concurrent connections without any endpoint latencies within the connections, other than those imposed by TCP.

(iii) The *block-sequential* connection structure model

We developed the *block-sequential* model, shown in Fig. 3.30, as another variation of the *Harpoon* model. In this model, all the TCP connections in the experiment behave like sequential connections but with only one epoch and no endpoint latencies within the connections, other than those imposed by TCP. Unlike the *Harpoon* and the *block-concurrent* models, however, this model introduces sequentiality and an inherent synchronization within a TCP connection. After connection establishment, the connection initiator sends one block, a $(a=a1+a2+a3)$ bytes in size, and upon receiving this *request*, the connection acceptor sends its *response* in one block, b $(b=b1+b2+b3)$ bytes in size. Thus all connections in this model are represented as single-epoch sequential connections, regardless of connection size.

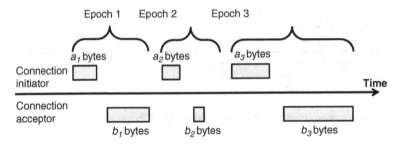

Fig. 3.31 The *a-b* connection structure model for sequential TCP connections

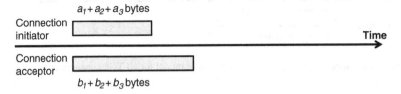

Fig. 3.32 The *a-b* connection structure model for concurrent TCP connections

(iv) The *a-b* connection structure model

The three connection structure models, discussed so far, model a connection based on its size alone. There were no endpoint latencies within the connections, and only the synchronization latency implicitly introduced by the request-response nature of the *block-sequential* model. In the *a-b* model shown in Fig. 3.31, we introduce the difference between sequential and concurrent connections, while still not including any measured endpoint latencies in the model. We do this by introducing the concept of epochs in sequential connections.

So, in this model, the original sequential connections replay in a pattern of request-response exchanges without the endpoint latencies representing processing times or other end system delays. For the original concurrent connections, the *a-b* model adopts the same representation as the *block-concurrent* model. This is because, in the absence of endpoint latencies, each endpoint of a concurrent connection will simply send its *a1*, *a2* and *a3* or *b1*, *b2*, and *b3* bytes in single blocks of size *a* and *b* respectively. This is shown in Fig. 3.32.

(v) The *a-t-b* connection structure model

This is the first connection structure model in which we explicitly introduce the endpoint latencies. We developed this *a-t-b* model mainly to differentiate between the effects of intra-epoch latencies and inter-epoch latencies. For the sequential connection, this model represents each epoch similar to the *a-b* model with the additional *intra-epoch* latency between the request and its response, as shown in Fig. 3.33.

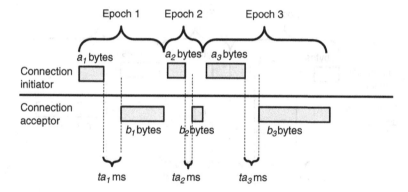

Fig. 3.33 The *a-t-b* connection structure model for sequential TCP connections

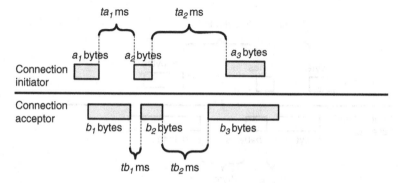

Fig. 3.34 The *a-t-b* connection structure model for concurrent TCP connections

For concurrent connections, there is no difference among the endpoint latencies. Each endpoint latency is associated with sending the preceding ADU from the endpoint and then waiting for the duration of the endpoint latency before sending the next ADU from that endpoint. It is not associated with any request-response exchange. Hence for concurrent connections, as shown in Fig. 3.34, we represent the connection using all the ADUs and the endpoint latencies as measured (similar to the *a-t-b-t* connection structure).

(vi)   The *a-t-b-t* connection structure model

All TCP connections in this model are represented using the same concepts originally developed by Hernandez-Campos et al. for the Tmix traffic generation system. So the *a-t-b-t* model represents a TCP connection with all ADUs and endpoint latencies and preserves all sequences or epochs exactly as measured in the original trace. This model is shown for the sequential and concurrent connections in Figs. 3.35 and 3.36 respectively. Note that concurrent connections have the same structure in both *a-t-b* and *a-t-b-t* models.

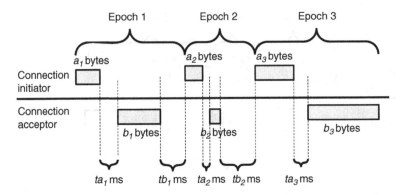

Fig. 3.35 The *a-t-b-t* connection structure model for sequential TCP connections

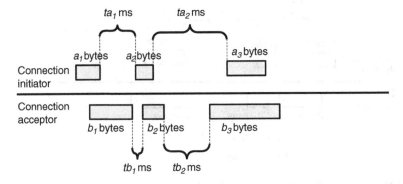

Fig. 3.36 The *a-t-b-t* connection structure model for concurrent TCP connections

3.3.1 Choice of Four Connection Structure Models

From the six different connection structure models described above, we chose to use only four among these to emulate the traffic for our complete sets of experiments. Our goal in picking the models was the following: introduce, one at a time, the following concepts within TCP connection structure modeling: size of the connection, client-server behavior between the two endpoints, the request-response exchange or epoch behavior between the client and the server, the fundamental dichotomy in application-level behavior that distinguishes connections as sequential or concurrent, and finally the endpoint latencies that represent intra-epoch and inter-epoch latencies in sequential connections or quiet times between sending of application data units in concurrent connections.

Here's why we chose (or did not choose) each of these connection structure models for our complete experimentation set.

Harpoon: This model inspired our development of the two block models. However, we did not choose this specific model because in faithfully adhering to the original *Harpoon* method of traffic generation, we had to model every TCP connection as two connections in the experiment. This led to difficulties in comparing performance metrics among the different models. The *block-concurrent* model is, therefore, a better representation of TCP connections for *Harpoon*-like traffic generation.

block-concurrent: We chose this model as it best represented the *Harpoon* model while also being the simplest model for emulating connection structure in terms of its size alone.

block-sequential: We chose this model as it introduced the notion of a client-server with inherent request-response synchronization while still preserving the simplest representation of a TCP connection by its size alone.

a-b model: This model was chosen because it introduces the concept of epoch structure within the sequential connections. Thus while there is the implicit addition of a time component to the structure in the synchronization implied by request-response exchanges, this model still does not explicitly include any of the measured endpoint latencies within the connections.

a-t-b model: We developed this model to differentiate between the effects of intra-epoch and inter-epoch latencies on the performance metrics. However, our preliminary investigations found that this model does not have significantly different effects from that of the *a-b* model. Here's why: the bulk (84%) of all intra-epoch latencies are below 500 *ms* for both UNC and IBM traffic and hence are not emulated in our experiments as explained earlier. Experiments using this model did not serve the original purpose envisioned while developing this connection structure model.

a-t-b-t model: We chose this model as it is a complete representation of connection structure for a TCP connection. To the *a-b* model, this adds all the endpoint latencies for both sequential and concurrent connections, thus explicitly introducing the dimension of time within a TCP connection.

3.4   Variations in Emulating Network Path Characteristics

Using the Tmix traffic generation system as the basis for generating traffic for all our experiments, we varied the emulation of the network path characteristics to study the effects of connection round trip times (RTT) on various metrics of performance. We developed seven different (some just subtly different) methods of RTT emulation. For our spectrum of RTT models, we have on one end the *nodelay* model where we completely eliminate the emulation of connection RTT, and on the other end the *usernet* model from Tmix where we emulate the specific minimum RTT for each connection as measured by analyzing the TCP/IP header traces from the captured traffic. We briefly discuss each of these models below.

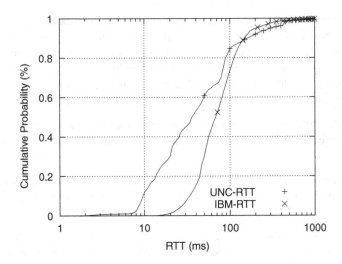

Fig. 3.37 CDF of round trip times (UNC and IBM traffic)

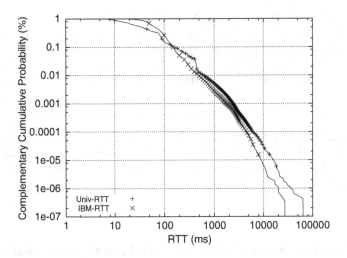

Fig. 3.38 CCDF of round trip times (UNC and IBM traffic)

In Figs. 3.37 and 3.38, we show the CDF and CCDF respectively for the minimum RTTs for connections in the UNC and IBM traffic. These figures are the same as Figs. 3.23 and 3.24. For six of our seven RTT models, we retained some measure of realism for RTT emulation, the only exception being the *nodelay* model. For five of the other six models, we used the empirical data shown in the RTT distributions above for each of the two input traces. Only the *10pathRTT* model was not derived from the above empirical distribution as explained below.

Tmix uses a modified version of *dummynet* that implements a user-level interface that can be used by Tmix instances to assign per-connection delays from the
input set of connection vectors. Although RTT is propagation delay between sender
and receiver, and in most cases the latency was emulated half on sender and half on
receiver, in the case of uniform RTT, the latency was emulated in only one
direction.

(i)   The *nodelay* model

First, the *nodelay* model was chosen simply as an extreme case to study why it
was important to model any form of RTT emulation rather than not model RTT
at all. For experiments using this RTT model, we replayed connections without
any round trip time latency. Thus the sending of packets within a connection
was still limited by the size of the connection, and the receiver window size,
but the round trip time experienced by the connections was on average only
1.42 *ms* with a standard deviation of 1.14 *ms*. This average was simply the
latency introduced by our laboratory network setup.

(ii)   The *meanRTT* model

For round trip time emulation in experiments using the *meanRTT* model, we
assigned a minimum RTT of 80 *ms* for all connections using the UNC trace and
92 *ms* for all connections using the IBM trace. These numbers were the measured average connection RTTs from the empirical distributions for these two
traces. This model was inspired by several leading studies, including the paper
by Sommers and Barford in 2004, describing the *Harpoon* traffic generation
system. It is worth noting here that this emulation effectively models a single
end-to-end path for all the millions of connections that play during the hour
long experiment. What do we mean? Note that when we assign different round
trip times to different connections in the laboratory network, we effectively
enable the emulation of different end-to-end paths (by assigning different
delays) for these TCP connections. But in the *meanRTT* model, we assign all
connections the same RTT value, thus reducing the experimental network to
emulate a single end-to-end path for all connections.

(iii)   The *medianRTT* model

This model of RTT emulation is very similar to the *meanRTT* model, creating
one shared end-to-end network path for all connections in an experiment.
Assigning the mean RTT of the distribution seems to be a more popular method
adopted in networking research, for example in [4]. However, studying the
traffic characteristics of traces captured on production network links shows
that a small fraction of connections with very long RTTs often skew the average RTT for the distribution. Hence although still an empirically derived value,
the mean RTT is less representative of the distribution of RTTs than the median
RTT. For example, the mean RTT for the UNC trace is 80 *ms* while its median
is only 36 *ms*. Similarly, the mean RTT for the IBM trace is 92 *ms* while its
median is only 68 *ms*.

Besides emulating a single shared end-to-end path for all connections in the experiment, assigning a single RTT value for all connections also significantly changes the traffic characteristics of the replayed trace. For example, with the *medianRTT* model, all those connections that had less than *medianRTT* in the original trace now take much longer to replay. Similarly, all those connections that had more than *medianRTT* delay in the original trace now replay faster. This has implications for several performance metrics as we show in Chapter 5.

(iv)   The *10pathRTT* model

All three models discussed so far – *nodelay*, *meanRTT*, and *medianRTT* – emulate a single shared path in the network for all connections in the experiment. The *10pathRTT* model expands the modeled network paths to a total of 10 different end-to-end paths for the connections in the experiment. The values chosen for these 10 paths were selected as follows: the TMRG common TCP evaluation suite [5] recommended 9 RTT path values based on some empirical measures. To this set of discrete values, we added a tenth RTT value to create our *10pathRTT* model.

Here's the small and discrete set of values that constitute the *10pathRTT* model: [4, 16, 28, 54, 74, 98, 124, 150, 174, 200] milliseconds. This set is used for both the UNC and IBM experiments.

(v)   The Discrete Approximation (*DA*) RTT model

We created this model from the empirical distribution of RTTs for the original trace. Hence the set of RTT values were different for the two traces – UNC and IBM. Our laboratory network has 30 pairs of traffic generators; hence we chose 30 values, thus creating 30 end-to-end paths in the network. The goal behind developing this model was to create as close an approximation of the empirical distribution of RTTs seen in the original trace as possible. For this we use the concept of a quantile function. A quantile function of a probability distribution is the inverse F^{-1} of its cumulative distribution function. Hence the quantile function returns the value of x such that $F(x) = P(X \leq x) = p$.

Our method of approximating the CDF of the RTTs was as follows: first we approximated the distribution such that we cut off the bottom 1% and top 1% of RTT values. These represented only 2% of connections but were skewing our overall approximations such that a very large portion of RTTs would be much larger than the median (or mean) RTTs. Now, with the remaining 98% of the distribution, we divided this distribution into 30 equal size bins, and then found the average RTT for each of these 30 bins in the distribution.

The resulting RTT values for UNC formed this set: [8, 8, 10, 10, 12, 14, 14, 16, 18, 20, 22, 24, 26, 30, 34, 38, 42, 48, 52, 60, 74, 80, 82, 86, 92, 98, 124, 172, 258, 420] milliseconds. The resulting RTT values for IBM formed this set: [22, 28, 32, 36, 40, 44, 46, 46, 48, 52, 54, 56, 58, 62, 66, 70, 74, 78, 82, 86, 92, 96, 102, 108, 114, 122, 136, 154, 188, 310] milliseconds.

(vi) The *uniformRTT* model

With the *uniformRTT* model, we made two significant changes to the assignment of connection RTTs discussed so far. First, instead of assigning specific delays to a small set of end-to-end paths, this model assigns a specific delay to each TCP connection. Thus instead of emulating 1, 10, or 30 shared end-to-end network paths, this model effectively enables emulation of a distinct end-to-end path for each TCP connection in the experiment. Second, the RTT values assigned to the connections were sampled from a discrete uniform distribution such that they approximately represented the middle 80% of the original RTT distribution for each trace. Hence for all experiments using the UNC trace, we sampled from the uniform distribution U[10, 200] milliseconds, and for all experiments using the IBM trace, we sampled from the uniform distribution U[30, 150] milliseconds.

(vii) The *usernetRTT* model

The *usernet* RTT model is adopted directly from the original design for RTT emulation used in the Tmix traffic generation system. In this model, every one of the millions of connections in an experiment is assigned the specific minimum RTT that was measured for that connection from analyzing the TCP/IP headers of the original trace. The complete distribution of RTTs used in this model is shown in Figs. 3.4.1 and 3.4.2.

3.4.1   Choice of Three RTT Emulation Models

From the seven different RTT emulation models described above, we chose to run complete sets of experiments using only three models. We have presented the results for a subset of experiments using the other four models in Chapter 6. Our goal in picking the three RTT models was the following: pick one model that emulates a single end-to-end path for all flows, pick one model that emulates a multiple but small set of end-to-end paths, and pick one model that creates the most faithful representation of the path characteristics of the original trace. Here's why we chose (or did not choose) each of these models.

nodelay: We did not choose this model for our full set of experiments. This model was used for preliminary experiments, simply to study the huge difference in performance metrics between not implementing any delay model, and implementing even the simplest model of RTT.

meanRTT: We chose this model for all our experiments because it is used in leading publications of networking research, for example in [4].

medianRTT: We decided not to use this model for our complete set of experiments. To create a single path for all connections, and given the distribution of RTTs, this model would actually make more sense since the mean skews the result in favor of the few large RTTs present in the distribution. However, since mean RTT is what is favored among networking researchers, we chose to evaluate using that model instead.

10pathRTT: We chose this model as it best satisfied our dual goals of using one multi-path RTT model which is also recommended by other networking researchers [5] as a model for all experimentation.

DA RTT: We chose not to use this model for two reasons. First, the *10pathRTT* model already satisfied our multi-path model requirement. Second, we discovered during our preliminary investigations that this model produces results very similar to the complete *usernet* RTT model because this model is the closest approximation of the empirical RTT distribution. Hence although we did not use it for our full set of experiments, we show some results with this RTT in Chapter 6, where we discuss some additional and interesting results from our study.

uniformRTT: We chose not to run our complete set of experiments using this model for two reasons. First, the *usernet* model captures the per-connection assignment of RTTs that this model introduces. Second, the *10pathRTT* already models a uniform distribution although with a much smaller set of values.

usernet: We chose this model to study the most precise emulation of RTT for empirically-derived, realistic traffic generation, where every connection is assigned its originally measured RTT value.

References

1. Comer D (2008). Talk titled "Lesson Learned from the Internet Project" given at the Department of Computer Science, University of North Carolina at Chapel Hill, October 2008.
2. Dag: 4.3 s single channel network monitoring card. http://www.endace.com/dag-4.3s-datasheet.html. Accessed 21 June 2011.
3. Hernandez-Campos F (2006) Generation and Validation of Empirically-Derived TCP Application Workloads. Dissertation, University of North Carolina at Chapel Hill.
4. Sommers J, Barford P (2004) Self-configuring network traffic generation. Proceedings of The Internet Measurement Conference
5. TMRG: The Transport Modeling Research Group. http://trac.tools.ietf.org/group/irtf/trac/wiki/tmrg. Accessed 9 July 2010.
6. Vishwanath KV, Vahdat A (2009) Swing: Realistic and responsive network traffic generation. IEEE/ACM Transactions on Networking

Chapter 4
Experimental Methodology

> A **theory** is something nobody believes, except the person who
> made it. An **experiment** is something everybody believes, except
> the person who made it.
>
> Albert Einstein

Experimental methodology plays an important role in protocol evaluations in net-working research. For experiments run in a laboratory network, as we did in this study, this methodology consists of the design of the network testbed, the calibration of the testbed components, and the design and running of experiments to test the hypotheses of the study. In this chapter, we first describe the methodology used for all experiments in this study. Next, using our control set for traffic generation comprising the *a-t-b-t* connection structure model and the *usernet* RTT model, we introduce the measurement and evaluation methodology that we use to run all experiments in this study.

What is this *control set* for traffic generation? We refer to the combination of the *a-t-b-t* model for connection structure and the *usernet* model for RTT as our *control set*. Here's why. In this study, we develop several new models for both connection structure and RTT emulation. The ideal method for comparing the effects of different models of traffic generation would be to compare the results for these models with the original traffic itself. That is, the real gold standard is obviously the original traffic captured on the production link. However, there are some differences between the original traffic and what is ultimately in the complete set of traffic components that we use as input in our experiments.

Now, Hernandez-Campos et al. have already shown that the Tmix models for connection structure (*a-t-b-t*) and network characteristics (*usernet* RTT, window size) can emulate any given input traffic in a realistic, reliable, and reproducible manner. That is, the traffic characteristics produced using the Tmix model at the packet level and byte level on the laboratory network link are the same as the traffic characteristics of the original input traffic to the Tmix system. Hence we use the Tmix models as our control set and compare all other models against them. As our

J. Aikat et al., *The Effects of Traffic Structure on Application and Network Performance*,   69
DOI 10.1007/978-1-4614-1848-1_4, © Springer Science+Business Media New York 2013

results bear out, this combination of models is indeed an excellent choice as a control for realistic traffic generation.

The rest of this chapter is organized as follows: in Sections 4.1, we describe the network configuration in detail. In Section 4.2 we discuss the process we used to calibrate the network, and its individual components, and present results from calibration experiments. Then, in Section 4.3, we describe our experimental procedures used in this study. And in Section 4.4, we introduce our control set for traffic generation. In Sections 4.5 and 4.6 we present the results for experiments using the control set in *unconstrained* and *constrained* modes.

The *unconstrained* mode is one in which the router-to-router link in the network is set to 1Gbps. In the *constrained* mode, that link is set such to 105 % of the expected average offered load on that link. More specifically, we recall that the average offered load for the UNC and IBM traffic is 471 Mbps and 404 Mbps respectively, on the high throughput or forward path on this link. Hence for experiments using the UNC traffic, we set this router-to-router link at 496 Mbps to create the *constrained* network mode. And for experiments using the IBM traffic, we set this link at 424 Mbps to create the *constrained* network mode. This way, the generated traffic consumes, on average, 95 % of the link capacity.

4.1 Network Configuration

We setup a network consisting of 60 PCs configured as traffic generators, two FreeBSD routers and three monitors collecting data on 1Gbps and 10Gbps fiber links at different points in the core of the network. All systems are Intel-based machines that run FreeBSD. A schematic diagram for this network is shown in Fig. 4.1. The traffic generators have 1Gbps Intel Ethernet interfaces and are attached to 1Gbps ports on the Ethernet switches. The two routers each connect to a 10Gbps fiber switch port on these switches. The switches aggregate the traffic on each subnet

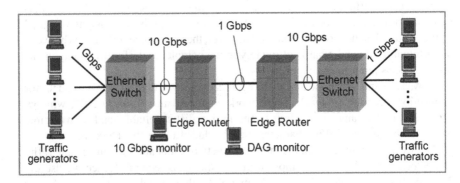

Fig. 4.1 Network Testbed for all experiments in this book

to a 10Gbps fiber connection to the router. The routers themselves are linked by a 1Gbps fiber link in the middle of the network. This is the link we refer to as the "router-to-router link" throughout this book. This is also the link that we manipulate to toggle the network environment between *unconstrained* and *constrained* modes for different experiments.

This network emulates a peering point between two ISPs with traffic flowing in both directions on the link between the two routers. During each experiment, traffic generated on the 30 traffic generators on each end is aggregated at the switches. This aggregate traffic then traverses the 10Gbps link to the router. The router on each end forwards the packets to the other side of the network. We capture this traffic as it traverses the router-to-router link. This physical network has a simple dumbbell topology. Logically, however, our traffic generation includes emulating per-flow minimum round-trip-times (RTTs). These minimum RTTs are obtained from a production network link on the Internet. This makes the network and the traffic traversing it effectively able to emulate a wide-area network.

4.1.1 Traffic Generators

Each subnet at the end of this dumbbell contains 30 PCs that serve as both traffic generators and data collection tools. These PCs range in capabilities from 450 MHz to 3 GHz in processing speeds, and 256 MB to 1 GB in memory. In each experiment, these traffic generators create application workloads and network characteristics based on the connection structure and RTT models used in that experiment. For all the experiments discussed in this study, unless otherwise specified, we assigned to each side of every TCP connection the exact maximum receiver window size that was determined from analysis of the original packet header trace. Connection durations and response times were measured and recorded by the traffic generators on each edge of the network during every experiment.

4.1.2 Routers

The two routers running FreeBSD are 3.6 GHz machines with 2 GB of memory. They are running the OpenBSD firewall software application known as packet filter (pf), which is a complete, full-featured firewall that has optional support for queuing. We use this packet filter module to restrict the bandwidth on the router-to-router link to desired limits during our experiments, and also to provide specific queue limits at the router's outgoing link. For experiments in the *unconstrained* network environment, we leave this 1Gbps router-to-router link unrestricted. This 1Gbps link capacity is significantly greater than the load generated from the two input traffic mixes we use in this study. For experiments in the *constrained* network environment, we restrict the router-to-router link to 105 % of the expected average

offered load. Hence, we set the router-to-router link to 496 Mbps for the UNC
replays and 424 Mbps for the IBM replays.

During calibration, we connected the two routers using either 1Gbps or 10Gbps
network interface cards. For all our experiments, however, we used only the 1Gbps
network interface cards to connect the routers. In all cases the router queues were
set to a large size (65,000 packets) which was determined to be sufficient to avoid
any packet drops at the queue so that loss rates were not a factor in any of the
results, even in *constrained* mode. We made this deliberate decision to provide such
a long queue so that there would be no losses in the network. We designed our
experiments to study the different effects on router queue dynamics due to different
models used for generating traffic. Providing a shorter queue and thus inducing
losses was out of scope for this study.

4.1.3 Monitors

We used two slightly differently monitoring and measurement configurations in the
network for calibrations versus the main set of experiments. In this section, we discuss the details of these setups and the reasoning behind the two different
configurations. Our main monitoring machine is a 3 GHz server class PC with 4 GB
of memory and running FreeBSD. For calibration, this machine was equipped with
a specialized traffic capturing card capable of collecting traffic up to 1Gbps load
between the two routers. The traffic capturing card is an Endace Systems' DAG
4.3S single channel network monitoring card. DAG technology provides 100 %
capture into host memory at full line rate for all packets on the link [1]. The traffic
captured by the monitor was analyzed using dagtools, and several diagnostic and
other tools developed at UNC, including an enhanced *tcpdump* program.

The trace collection process in the laboratory is similar to the trace collection
process on any production link. Only the packet protocol headers (IP and TCP) are
collected, and the timestamp of the packet arrival is recorded. For all calibration, we
use the specialized DAG hardware to extract headers and provide accurate time-
stamps. The DAG trace collection has accuracy in the order of nanoseconds for
timestamping of the packets. Such accurate packet header traffic captures were
essential for calibration and testing so that we could verify that the connection structure and RTT models were being emulated exactly as designed.

Once the laboratory network was calibrated, we changed the monitoring setup for
all experiments as follows. We used three FreeBSD machines for monitoring and
measurement. The first machine is a 2.3 GHz machine with 2 GB of memory, the
second is a 1.5 GHz machine with 512 MB of memory, and the third is a 3 GHz
machine with 4 GB of memory. The first two recorded traffic data traversing the
router-to-router link in both directions, one recording counts of the bytes and packets in hundred microsecond intervals, and the other recording all SYN, FIN, or RST
packets to count active connections in the network. The third monitor recorded, in
hundred microsecond intervals, the arrival of bytes and packets to the router queue.

Both our input traffic sets – UNC and IBM – had loads that were not symmetrical in the two directions. For queue lengths, we were therefore interested only in the router queue on the high throughput path of this traffic. Hence the third machine monitored the 10Gbps fiber link aggregating the traffic between the switch and the router only on the path of this higher traffic throughput. At the router we recorded a log of the queue size (number of packets in the queue) sampled every 10 milliseconds.

The two switches in the core of the network are 26-port HP Procurve 3400 cl switches, each connected to a 48-port Netgear GS748T switch. Each HP switch has 24 1Gbps copper ports and two 10Gbps fiber ports. Each Netgear switch has 48 ports which can be configured as 40 ports of 1Gbps copper and eight ports of 1Gbps fiber. In order to avoid any bottleneck on the switch connections between the Netgear and the HP switches, we setup a 4Gbps trunk between each pair of switches. This trunking is based on the IEEE 802.3ad Link Aggregation Control Protocol (LACP). This is an IEEE standard for link aggregation supported by both sets of switches (HP and Netgear). Such a setup enables a virtual link of 4Gbps between the switches. Key features of link aggregation are: it is performed above the MAC layer, it assumes all links are full-duplex and same data rate, traffic is distributed packet by packet, and all packets associated with a given flow are transmitted on the same physical link to prevent mis-ordering of packets.

4.2   Network Calibration

Once we have configured the network, it must be calibrated before any experiments can be reliably run using this network. But why do we calibrate a network? The main motivation for network calibration is to ensure that the network, or any of its individual components, do not present any resource constraints (unless otherwise designed to do so, as in a bandwidth *constrained* link) when running experiments. The way we verify this is through calibration. Calibration involves first identifying the set of all inputs to the experiment, deciding what the outputs will be, and figuring out the correlations, if any, between these inputs and outputs.

The goal of calibration then is to ensure that these correlations are not influenced by an unintended lack of resources in the network. Consider the case where the throughput in the core of the network (output metric) is dependent on the number of TCP connections (input variable) in the traffic. If increasing the number of TCP connections linearly increased the link throughput in the core up to a certain point, then we could use this correlation to calibrate the network and determine the reliable working range of inputs and corresponding outputs for which this relationship holds. Say, for the sake of simplicity, that each TCP connection generated 1 Mbps of traffic, and each traffic generator could handle 100 such connections without overloading any resources on these machines. With 30 such traffic generators, we could then easily generate 3Gbps of traffic into the network. Assume that the traffic generators have 1Gbps link each, and all the aggregation links are 10Gbps. What if one of the routers in the network were continuously overloaded with 100 % CPU utilization trying to forward packets at this rate of 3Gbps? The router would start dropping packets and this affects the previously established correlation between number of TCP connections and the throughput in the network. This is a case where lack of resources at one point in the network affects the input-output dynamics of the experiment.

During calibration, we push the network components, one at a time to determine its limits. Then we design our experiments so that each network component is working well below its resource limits. Hence we calibrate the network by designing and running a set of experiments that stress-test every component of the laboratory testbed system with the goal of ensuring that no single network component (individually or as part of the full network), presents a resource bottleneck for the main set of experiments designed to test the hypotheses of this study. Toward this end, we designed a series of calibration experiments with target loads of bytes and packets that were much higher than the target loads in any of the main set of experiments of this study. If these higher target loads were achieved, then these experiments would ensure that the traffic generators, routers and monitors would not present any bottleneck in the main set of experiments.

4.2.1 Calibrating Routers

The two routers in the core of the network forward packets, constrain the router-to-router link to operate at a specified bandwidth (by managing an outbound queue of packets to this link), and collect measurement data. To calibrate the routers we had to ensure that their CPU utilization was acceptable when performing all of these tasks in any given experiment. The maximum average offered load for any experiments in this study is 471 Mbps in one direction. Hence, to stress test the routers, we designed two sets of experiments. The first set used the *iperf* program [3] between two pairs of traffic generators using four TCP connections to generate an aggregate load that was only limited by a constraint of 622 Mbps imposed on the forward path link between the routers. The traffic on the reverse path was about 550 Mbps. The link was *constrained* on the forward path to create a worst-case scenario to stress-test the routers. That is, the router had to be able to forward packets onto the *constrained* link at the rate of 622 Mbps while also managing the outbound queue of packets to this link.

The second set used the Tmix traffic generation system between 32 pairs of traffic generators using 8.5 million TCP connections to generate 740 Mbps in the forward path and 230 Mbps in the reverse path. The two routers were determined not to be a bottleneck at any of these high loads of traffic. That is, the routers were able to forward packets at these rates without dropping packets, and do so while maintaining an acceptable level of CPU utilization, that is, at or below 95 % utilization at all times.

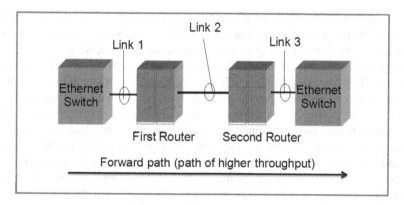

Fig. 4.2 Routers' inbound and outbound links

We also experimented with different clock frequencies on the routers setting them at 100 Hz, 1 KHz and 10 KHz. At 1 KHz, the clock interrupts occur 1000 times a second. This is the frequency at which all the traffic generation systems operate. We wanted a higher frequency of clock interrupts on the routers to allow for a finer granularity for timers. Higher frequencies, however, also cause processing overhead. Hence we ran experiments with different clock frequencies to study the balance between these two tradeoffs of finer timer granularity versus higher CPU utilization. We found that the 10 KHz clock frequency resulted in slightly higher but still well below 90 % utilization at all times, for the throughput levels designed for our experiments. Hence we used 10 KHz as clock frequency for our routers in all our experiments.

We ran another set of calibration experiments to test the following: the CPU utilization on the routers seemed dependent on the inbound and outbound links on these routers. Let us discuss this using Fig. 4.2 shown below. From the figure we have the following: for the forward path or higher throughput path, the first router's inbound and outbound links are labeled "link1" and "link2" respectively. Similarly, the second router's inbound and outbound links are labeled "link2" and "link3" respectively for the forward or higher throughput path.

Through initial calibration, we had found that the second router on the forward path showed higher CPU utilization than the first router on that path. This seemed counter-intuitive at first. But we determined through a series of specially designed experiments that this was due to more efficient processing of incoming packets on the first router's 10Gbps inbound NIC than the second router's 1Gbps inbound NIC for the traffic on the forward path. We conjecture that this is a difference in the efficiency of the drivers for the two network interfaces though they are both Intel network cards. We verified this by running several experiments with varying loads using 1Gbps NICs throughout, and then repeating these experiments with 10Gbps NICs throughout as well as combinations of 1Gbps and 10Gbps links.

In the presence of a 10Gbps NIC on the second router for inbound traffic on the forward path, this second router dropped its CPU utilization to the same lower levels as that of the first router. For all our experiments, however, we used the 1Gbps router-to-router link and 10Gbps link from the switch to the router after determining that the slightly higher router CPU utilization on the second router did not present a bottleneck for the traffic. That is, although this second router had significantly higher CPU utilizations with this configuration (see Fig. 4.2 (b)), the level of CPU utilization achieved for the throughputs at which we were operating in our experiments was acceptable. That is, we found the router utilization to be below 80 % for the middle 40 minutes in all our experiments. Note that we report performance results using only the data from the middle 40 minutes of each experiment. In the set of iperf calibration experiments presented below, we used the worst case (in terms of testing CPU utilization) of having 1Gbps Intel NICs on both the routers on the incoming and outgoing paths.

We also ran some experiments to determine the appropriate size for the transmit buffer on the router's network interface card (NIC) driver. Here's why. When this transmit buffer on the NIC driver was left at its default value, there were times in an experiment when the router's outbound queue (managed by the pf module) seemed to drain; that is the router queue had no packets in it. However, the corresponding queuing delay results did not support this apparent draining of the queue. Further investigation revealed that these packets that were dequeued from the router's outbound queue were actually being enqueued in the NIC's onboard transmit buffer before being transmitted out on the link. We then ran experiments with different buffer sizes for that transmit buffer to determine an optimum size that would be small enough not to cause noticeable additional queuing delays but also large enough not to drop packets. We found this number to be 4 packets instead of the default 256 packets for the transmit queue.

4.2.1.1 Iperf Experiments for Calibrating Routers

Iperf, orginally developed by NLANR (National Laboratory for Applied Network Research), is a tool often used by networking researchers for some basic measurement and testing in a network. This includes testing of bandwidth, latency, jitter and loss using TCP and UDP flows. Iperf allows the tuning of various parameters and UDP characteristics. Iperf uses FTP-like data streams. The throughput of each TCP connection is, therefore, mostly dependent on the receiver window size and available network bandwidth. The data is sent only in one direction for each connection, with pure acks traversing the opposite direction.

For calibration, we ran several experiments using iperf: first, between two pairs, and then four pairs, of traffic generators to generate TCP traffic in both directions. We ran every experiment in the forward and reverse directions to ensure there was no difference in the setup of the two routers.

In these experiments, we use the term *forward* to refer to the direction in which there is higher throughput of bytes and packets in the network. We refer to the

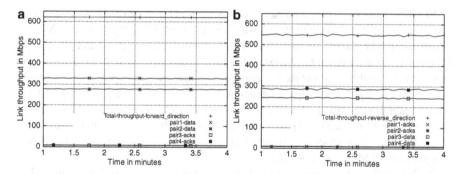

Fig. 4.2 (a): Throughput for the iperf flows – forward direction (b): Throughput for the iperf flows – reverse direction

opposite direction as the *reverse* direction. To run the experiments, we used four pairs of machines with two pairs (pair1 and pair2) sending data in the forward direction using one TCP flow each and two pairs (pair3 and pair4) sending data in the reverse direction using one TCP flow each. Hence the forward path also carried the acks for the TCP connections between the traffic generators in pair3 and pair4 while the reverse path carried the acks for the TCP connections between the machines in pair1 and pair2. Thus we had two TCP flows in the forward path with their corresponding ack flows in the reverse path, and we had two TCP flows in the reverse path with their corresponding ack flows in the forward path.

Each iperf experiment ran for five minutes. We collected data on the iperf clients and servers, the routers and at the router-to-router link using the monitor with the DAG capture card. We ran these iperf experiments at different loads *constrained* by the capacity of the router-to-router link. We set this link to 100Mbps, 200Mbps, and so on up to the *unconstrained* mode of 1Gbps. Recall that since iperf can generate connections with unlimited data, these TCP flows grow their window size up to the maximum available bandwidth. The TCP receiver maximum windows were set to 64 KB. And since we provided very large router queues, the packets were queued without any packet loss at the *constrained* link.

As we explained earlier, the *constrained* experiments represent the maximum usage of resources on the routers because the routers have to forward packets onto the *constrained* link while also managing the outbound queue of packets to this link. Hence, we show the throughput results for the worst-case experiment in the above mentioned series of iperf experiments. We only show the results for the middle 3 minutes of that 5 minute experiment in Figs. 4.2 (a) and (b) since this is the stable region. In this experiment, we emulated connection RTTs by using *dummynet* to set delays of 10 ms and 15 ms on the iperf servers, thus delaying all acks going from server to client (Iperf sends data from client to server). Figure 4.2 (a) shows the byte throughput in the forward (high throughput) direction and Fig. 4.2 (b) shows the throughput in the reverse direction. The router-to-router link was *constrained* at 622Mbps in both directions. Hence each figure shows the throughput in one direction. The throughput in each direction consists of two data streams and two acks streams.

Figure 4.2 (a) shows the two data streams for pair1 and pair2. These two flows had an average throughput in the forward direction of 332 Mbps and 275 Mbps, which along with the throughput of the two acks streams from pair3 and pair4 of 8 Mbps and 7Mbps, totaled 622Mbps or the full capacity of the link. Figure 4.2 (b) shows the throughputs of the data streams on the reverse paths (generated by pair3 and pair4) and the ack throughput generated by pair1 and pair2 on this path. The data throughputs on this path are slightly lower at 293 Mbps and 249 Mbps. This reflects the fact that pair3 and pair4 were the least capable machines in the network so the total on this path did not hit the link capacity limit. The ack throughputs on this path are 10 Mbps and 8 Mbps.

Each iperf flow shown above sends TCP data in only one direction with pure acks sent in the other direction. Hence, we note the following interesting data collected from these experiments. In the forward direction, 38 % of Ethernet frames were 66 bytes in size (acks for pair 3 and 4) and 62 % were 1514 bytes (data for pair 1 and 2). In the reverse direction, 43 % of Ethernet frames were 66 bytes in size (acks for pair 1 and 2) and 57 % were 1514 bytes (data for pair 3 and 4). There were 24.4 million packets in the forward direction and 23.6 million packets in the reverse direction, generating roughly 80 Kpps (thousand frames per second) in each direction during this short 5-minute experiment. Why does this matter? We also measured the CPU utilization at the routers to be 95 % and 85 % on average for the first and the second routers respectively. This difference in router CPU utilization for the same data being handled had to do with the fact that the network interface card handling this incoming traffic was 10Gbps on one router and 1Gbps on the other. Details of experiments specifically exploring this difference in CPU utilization were discussed in Section 4.2.1 (Calibrating Routers). Hence, we can conclude that the routers can handle packet throughputs of 80Kpps without presenting resource constraints on the routers. The main experiments of this study (presented in Chapter 5) were all designed for lower target loads (total byte and packet throughputs) than these.

4.2.1.2   SECTION III Tmix Experiments for Calibrating Routers

Iperf experiments use large size data packets, while most traffic on the Internet consists of a large variety of packet sizes. Hence we ran another set of calibration experiments using a captured UNC trace as input to the Tmix traffic generation system. This system consists of several components. The traffic generation tool, *Tmix*, replays the source-level behavior of a set of input connection vectors using real TCP sockets in a FreeBSD environment. [2]. *Usernet*, a modified version of dummynet, implements a user-level programming interface that is used by tmix instances on the traffic generators to assign per connection delays as specified in the input set of connection vectors.

Finally, a single program, *treplay*, is used to control the setup of the experimental environment, configure and start the tmix instances (assigning them a subset of connection vectors and traffic generation peer), and collect the results. Tmix instances rely on the standard socket interface to create a connection, send and receive ADUs,

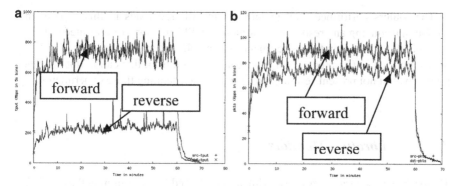

Fig. 4.3 : Throughput for Tmix calibration experiment

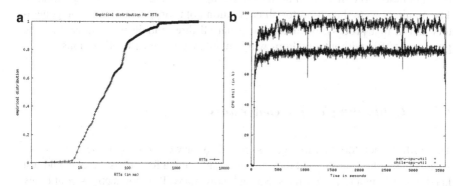

Fig. 4.4   (a): Distribution of RTTs (b) : Router CPU utilizations

and to close the connection. For every experiment, all the machines are first initial-
ized and configured. Then the routers and monitors start their monitoring programs
followed by the traffic generators running instances of the tmix program.

For the router calibration using Tmix, we tested the network using several different offered loads, scaling the same input traffic to these higher loads in each case,
using the block-resampling methods from Hernandez-Campos 2006. In this section,
we describe the experiment with the highest of these loads because they created the
most stress on the routers in the network. In this highest load case, we had an average byte throughput of 740Mbps in the forward direction and 230 Mbps in the
reverse direction. The corresponding average packet throughputs were 89Kpps in
the forward direction and 74Kpps in the reverse direction. We show the time series
of byte and packet throughputs for the entire hour long experiment in Fig. 4.3 (a)
and (b) respectively. In this experiment, we used 32 pairs of traffic generators, and
the average load per pair of traffic generators was 1.4 times the highest average load
per traffic generator in the experiments reported in Chapter 5. The distribution of
RTTs for the connections (shown in Fig. 4.4(a)) is similar to that of the UNC traffic
used for experiments reported in Chapter 5.

The routers performed well under the high offered loads in this experiment without introducing any resource constraints of CPU, memory, or allocated buffers. And the offered loads in this experiment were much higher than the loads in experiments reported in Chapter 5. Hence we can conclude from this calibration that the routers would not present resource constraints when running those experiments.

4.2.2 Calibrating Monitors

There were no separate sets of experiments conducted for determining the capabilities of the three monitors. However, buffers on the monitors were tuned during initial calibration to collect data at high throughputs for the hour long experiments. The monitors were then used for all calibration both for routers and traffic generators and in that process, we determined that all three monitors could capture the generated traffic without any drops while maintaining low CPU utilizations.

4.2.3 Calibrating Traffic Generators

The traffic generators had to be stress-tested to answer two main questions. First, what is the highest throughput they can generate using a few flows – this would test handling of byte and packet rates. Second, how many TCP connections could they manage while running Tmix? Running Tmix with a few thousand flows would test the CPU, memory and buffer management capabilities for managing these connections. Toward this end, we calibrated the traffic generators (similar to the router calibration) as follows. First, we generated a few high throughput TCP flows per traffic generator pair, sending large packets using the iperf program. Then we generated more than one hundred thousand TCP flows per traffic generator pair, sending a diverse mix of packet sizes and flow sizes using the Tmix traffic generation system.

The number of connections managed per traffic generator is an important factor in calibration. This is because with a few thousand TCP connections alive per second on average per traffic generator, the traffic generators must manage the CPU, memory and buffer resources to keep state for all these connections while servicing each connection in a round-robin fashion. For the calibration using iperf, we refer to Figs. 4.2 (a) and (b) back in Section 4.2.1. The two pairs of traffic generators used in that experiment represented the most capable and the least capable pair of machines in our set of 30 pairs of traffic generators (with respect to their processing and memory capabilities). Each of these four PCs served as either client or server, and thus generated two data streams and two ack streams in each direction. As seen in Figs. 4.2(a) and 4.2(b), the least capable of these traffic generators was able to generate iperf data traffic of at least 240Mbps. This is more than an order of magnitude higher throughput than what we require for the experiments reported in Chapter 5.

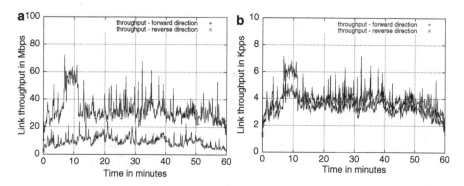

Fig. 4.5 Throughput for Tmix calibration experiment for least capable traffic generator pair

For all experiments reported in Chapter 5, we require each traffic generator to generate traffic that is less than 20Mbps. Also, the CPU utilizations on these PCs during the iperf experiments were quite low – they were less than 20 % on each traffic generator. Hence, this iperf experiment gave us an upper bound for the traffic generators in terms of the total throughput they could each generate using only one connection per pair even for the least capable traffic generators.

For traffic generators, generating and managing thousands of TCP connections over an hour long experiment is a better stress test than generating a few TCP flows of very high throughput. Hence we ran experiments using Tmix with realistic traffic captured at the UNC campus link. This input was an hour long trace captured on December 7, 2007 starting at 11:30 AM. This represents peak campus-Internet traffic just like the January 2008 trace we used for the experiments reported in Chapter 5. Unlike the router calibration using Tmix (where we ran all pairs of traffic generators at once), we ran these experiments using only one pair of machines at a time to determine their capability and find any bottlenecks. The median of the maximum CPU utilizations on the most capable and least capable pair of machines were 53 % and 72 % respectively.

In this section, we present the results only for the worst-case – that is, the least capable pair of machines running Tmix. As shown in Fig. 4.5 (a), the average throughput was 31Mbps and 11Mbps in the forward and reverse paths. The corresponding packet throughputs were 3.8 Kpps and 3.2 Kpps as shown in Fig. 4.5 (b). It must be noted that by generating only 30Mbps of traffic with each of the 30 pairs, we could potentially generate 900 Mbps using all 30 pairs, without these traffic generators presenting any bottlenecks. This is almost double the traffic generated in any of the experiments reported in Chapter 5, and thus provides a much higher upper limit for each traffic generator.

Our calibration for this least capable pair of traffic generators showed that even at these relatively high loads, they replay the input traffic using Tmix exactly as intended. We verified this as follows. We extracted the following data from the connection vectors representing the input traffic for this experiment: roundtrip times,

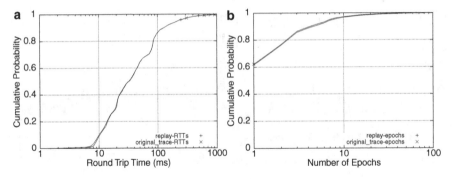

Fig. 4.6 (a) and (b): CDF and CCDF for input and output round trip times

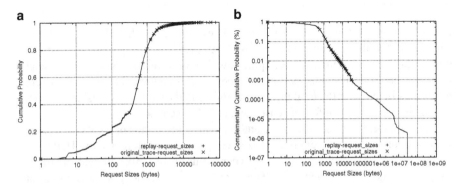

Fig. 4.7 (a) and (b): CDF and CCDF for input and output request sizes

number of epochs in sequential connections, request sizes and response sizes in sequential connections and the 'a' and 'b' ADU sizes for the concurrent connections. Then during the hour long experiment, we acquired the packet header trace of the generated traffic on the router-to-router link using the 1Gbps DAG card. We then processed and analyzed this trace for the same set of measures as we did for the input connection vectors (derived from the trace on the production link). Figures 4.6 through 4.10 compare the distributions of various measures of TCP connections in the original traffic (input to the traffic generators) and the results of the calibration replay experiments (output to the traffic generators).

Figures 4.6 (a) and (b) compare the CCDFs of the original and replay-generated distributions for connection RTTs and number of epochs per connection. We observe that these distributions match closely indicating that the traffic generator pair is replaying the traffic as designed. Similarly, Figures 4.7 (a) and (b) compare the original and replay-generated distributions for request sizes in sequential connections, showing the CDF and CCDF in the two figures respectively.

Figures 4.8 (a) and (b) confirm that this traffic generator pair also replayed response sizes in sequential connections as designed. Figure 4.9 (a) and (b), and

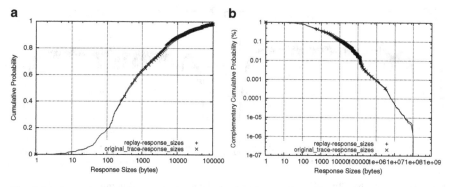

Fig. 4.8 (a) and (b): CDF and CCDF for input and output response sizes

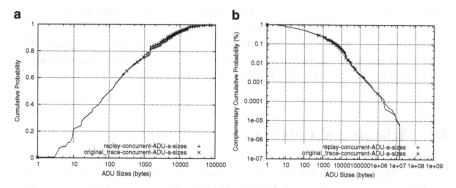

Fig. 4.9 (a) and (b): CDF and CCDF for input and output concurrent 'a' sizes

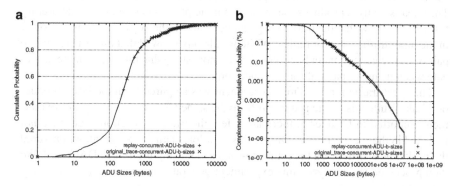

Fig. 4.10 (a) and (b): CDF and CCDF for input and output concurrent 'b' sizes

Fig. 4.10 (a) and (b) compare the original and replay-generated distributions for the ADUs in concurrent connections in the two directions for each connection. These are the *a* and *b* sizes as shown in these figures. Figures 4.11 (a) and (b) show that,

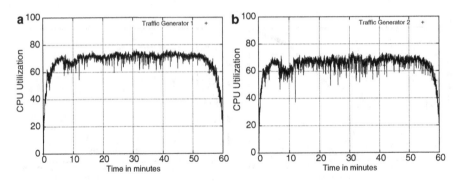

Fig. 4.11 (a) and (b): CPU utilization for the two traffic generators used in this experiment

throughout this experiment, the CPU utilizations over 1 second intervals were less than 75 % for the two traffic generators.

From this data, we concluded that the traffic generators would replay the traffic using the Tmix traffic generation system as designed, and no traffic generators would present a bottleneck in the experiments we report in Chapter 5 and 6.

4.3   Verification of Tmix Replay

In the previous sections, we discussed calibration of routers, monitors and traffic generators. Having completed calibrating the network, we now show that our full laboratory network testbed was configured properly to replay traffic using the Tmix traffic generation system for the experiments reported in Chapter 5. We verify that Tmix realistically reproduces the traffic from the production link in our laboratory testbed. We show that the traffic we generate bears all the key characteristics found in the input traffic used for replay. While we already showed that this is true for one pair of traffic generators in Section 4.2.3, we now show that this holds in the aggregate when using all pairs of traffic generators.

The input traffic for this Tmix experiment was acquired from the UNC production link. While this traffic is derived from the same UNC traffic we use for experiments reported in Chapter 5, there are a few thousand connections that were not included in these experiments. During this hour long Tmix experiment, we captured the packet header trace on the router-to-router link using the 1Gbps DAG card. We then processed and analyzed this experiment-generated trace for several key measures of traffic.

Figures 4.12 (a) and (b) show the throughput in Mbps and Kpps computed in 5 second intervals. The average byte throughput in the middle 40 minutes of this replay was 451 Mbps with a standard deviation of 35Mbps in the forward direction, and 165 Mbps with a standard deviation of 19 Mbps in the reverse direction. The corresponding average packet throughput was 56 Kpps with a standard deviation of

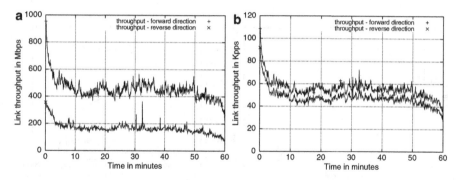

Fig. 4.12   Throughput for Tmix verification experiment

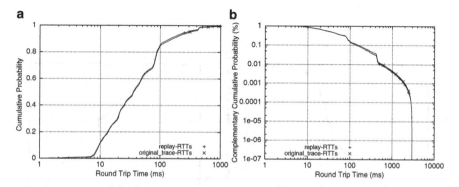

Fig. 4.13   (a) and (b): CDF and CCDF of connection RTTs for Tmix verification experiment

3 Kpps in the forward direction, and 47 Kpps with a standard deviation of 2.6 Kpps in the reverse direction.

As we observed in the router calibration using Tmix, we find there is a significant spike in throughput at the beginning of the replay due to the 30 pairs of traffic generators starting all at once, and all of them starting TCP connections in the first few minutes of the experiment. There is also a significant decay in throughput during the last few minutes of the experiment. For results reported here, we use data collected during minutes 10 to 50 of the replay.

We now verify this Tmix experiment (similar to Section 4.2.3) by visually comparing the distribution of several key measures of the traffic on the production link with the corresponding measures for this replay using the CDFs and CCDFs for these parameters. We extracted the following distributions from measurements of both sets of traffic: connection minimum RTTs, number of epochs in sequential connections, request sizes and response sizes in sequential connections, and the 'a' and 'b' ADU sizes for the concurrent connections.

Figures 4.13 through 4.22 show the distributions for each of these measures comparing data from the original trace (input to the experiment) with data from the

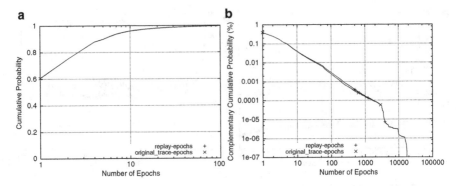

Fig. 4.14 (a) and (b): CDF and CCDF of number of epochs per connection for Tmix verification experiment

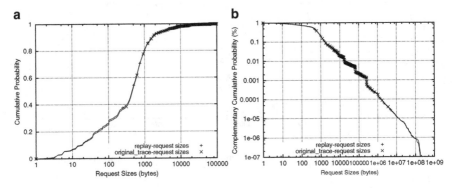

Fig. 4.15 (a) and (b): CDF and CCDF for request sizes for Tmix verification experiment

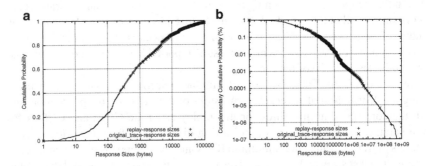

Fig. 4.16 (a) and (b): CDF and CCDF for response sizes for Tmix verification experiment

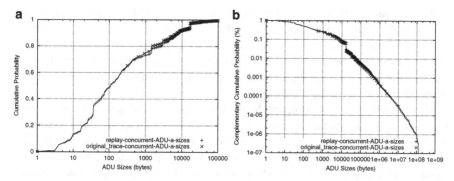

Fig. 4.17 (a) and (b): CDF and CCDF for concurrent 'a' sizes for Tmix verification experiment

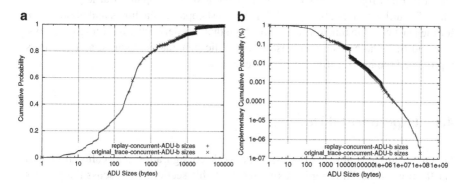

Fig. 4.18 (a) and (b): CDF and CCDF for concurrent 'b' sizes for Tmix verification experiment

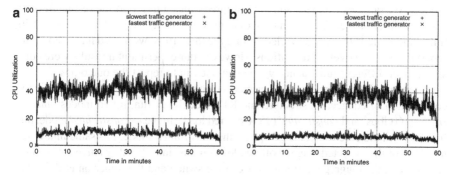

Fig. 4.20 (a) and (b): CPU utilization for the most and least capable traffic generator pairs on each subnet

replay experiment. Figures 4.14 (a) and (b) compare the distributions of the minimum round trip times per connection for the original trace and the replay. The two distributions match very closely showing that we emulated the connection RTTs exactly as required. Similarly Fig. 4.15 (a) and (b) compare the distributions for the number of epochs for sequential connections in the original trace with the number of sequential epochs in the replay.

Figures 4.16 and 4.17 compare the distributions of the measured request sizes and response sizes with the distribution of request sizes and response sizes produced by the replay experiment. Figures 4.18 and 4.19 compare the distributions of the 'a' sizes and 'b' sizes in concurrent ADUs with the corresponding distributions measured from the production link. As shown in all these figures, the replay trace has the same distributions of measures of connection structure and network characteristics (RTTs) as the original trace which was captured on the Internet link.

The CPU utilizations on the traffic generators were fairly low (see Figs. 4.20 (a) and (b)). Each figure shows the most capable and least capable traffic generators. The CPU utilization is about 10 % for the most capable machines and about 40 % for the least capable machines. We conclude that our network was configured properly and the whole system consisting of traffic generators, routers, and monitors reproduced the input traffic exactly as intended.

4.4   Experimental Design

So far, in this chapter, we discussed network configuration and calibration, and verification of the Tmix replay experiment. In this section, we discuss the process of developing the overall design of experiments to prove or disprove our hypotheses in this study. We conducted experiments using all combinations of the four connections structure models and three RTT emulation methods (described in Chapter 3). In Chapter 5, we report the results from combinations of experiments using these models. We ran every experiment at least three times, but report the results of only one experiment for each combination of connection structure model and round trip time emulation. If the results varied among the three runs, we would have chosen to report the average over all repetitions. However, our experimental results were consistent over different runs; hence we picked one run to report the outcomes.

We repeated the entire set of experiments using both UNC and IBM traffic. Every combination of connection structure and RTT model was run in two modes: *unconstrained* (1Gbps) and *constrained* (95 % offered load). In the *unconstrained* mode, the link between the core routers is 1Gbps. In the *constrained* mode, this same link is set to 105 % of the expected average offered load on this link. Whether *unconstrained* or *constrained*, the (aggregation) link between the switch and the router on each of the two subnets was always 10Gbps for all experiments. For experiments with UNC traffic, the average uncongested load was 471 Mbps and hence the *constrained* link capacity was set to 496 Mbps. For experiments with IBM traffic, the average uncongested load was 404 Mbps and hence the *constrained* link capacity was set to 424 Mbps.

For every experiment, we collected measurements at various points in the experimental network. We then analyzed these measurements to study the effect of connection structure models and round trip time emulation methods on four key performance metrics. These performance metrics are connection durations and response times (both recorded on the traffic generators for every TCP connection), the router queue length (recorded on the router for its outbound queue), and active connections (recorded on one of the two monitors on the router-to-router link).

Unlike the calibration experiments, we did not use the monitor with the DAG card in these experiments. Hence, we did not capture the packet header trace for all the traffic on the link. Instead, we measured throughput on the link, counting every byte and every packet traversing that link in 100 microsecond intervals. In this section, all figures show throughput results aggregated over 5 second intervals. The arrival of packets and bytes into the network is fairly bursty, representing the nature of arrivals onto the Internet link at which the original trace was measured. The aggregation uplink before the core routers is a 10Gbps link in our testbed network. On that link, we measured byte arrivals well over 1Gbps at sub-10 ms intervals. In the figures, 'Mbps' indicates throughput in units of Megabits per second, and 'Kpps' indicates throughput in packets with units of Kilopackets (thousands of packets) per second. Every experiment was run for 60 minutes, but all data shown in the results sections are for the middle 40 minutes to eliminate startup and termination effects. It was determined during calibration that allowing 10 minutes for startup effects to diminish and 10 minutes for termination effects to diminish was adequate to account for such effects.

4.4.1 The Control Set: a-t-b-t with usernet

In this section, using our control set for traffic generation comprising the *a-t-b-t* connection structure model and the *usernet* RTT model, we introduce the measurement and evaluation methodology that we use for all experiments reported in Chapter 5. As we explained earlier, we adopted the combination of the *a-t-b-t* model for connection structure and the *usernet* model for RTT as our *control set*. We use this set to compare the effects of different models of traffic generation on application-level and network-level performance metrics. While the real gold standard is obviously the original traffic captured on the production link, Hernandez-Campos et al. have already shown that the Tmix models for connection structure (*a-t-b-t*) and network characteristics (*usernet* RTT, window size) can emulate any given input traffic in a realistic, reliable, and reproducible manner. In Section 4.3, we successfully verified that the output characteristics of the traffic generated matched their corresponding input parameters for traffic generation, given our particular experimental setup. Hence we use the Tmix models as our control set and compare all other models against them. As our results bear out in this study, this combination of models is indeed a good choice as a control set for realistic traffic generation.

We have already used this control set of models for the three Tmix experiments presented for calibration and verification in this chapter so far. So what differentiates those experiments from the ones below? Those experiments used only UNC traffic, not IBM traffic. Though the traffic sets for those experiments were acquired from the UNC production link, they are different from the traffic set we use for results reported in Chapter 5.

4.5 a-t-b-t with usernet in Unconstrained Mode

In this section, we discuss the results for two experiments (one using UNC traffic, and the other using IBM traffic) modeled with the control set and run in the *unconstrained* network mode. We present results for the time series of throughput followed by results for the performance metrics: connection durations, response times, router queue length, and active connections. For all of these measures, we present results for both experiments, comparing them on the same figure wherever possible.

4.5.1 Throughput

Figures 4.21 and 4.22 show the byte throughput time-series for the experiments using the UNC and IBM traffic in the *unconstrained* mode. Figures 4.23 and 4.24 show the corresponding packet throughput time-series. We present the throughput time series because it is the most common and familiar measure of characterizing traffic on any production link or, in this case, traffic generated in the laboratory. These figures show throughput data averaged over 5 second intervals.

Figure 4.21 shows that the mean throughput for a replay of UNC traffic is 471 Mbps with a standard deviation of 34 Mbps. Figure 4.22 shows that the mean throughput for a replay of IBM traffic is 404 Mbps with a standard deviation of 37 Mbps. The corresponding packet throughputs are shown in Figs. 4.23 and 4.24. The mean packet throughput for a replay of UNC traffic is 60 Kpps with a standard deviation of 3.0 Kpps. And the mean packet throughput for a replay of IBM traffic is 62 Kpps with a standard deviation of 3.7 Kpps.

It is worth noting that the throughput time-series for the experiment using the UNC traffic is stationary for the hour. The throughput time-series for the experiment using the IBM traffic, however, is non-stationary. That is, for the experiment using the IBM traffic, the mean of the throughput changes significantly in the latter half of the time-series (see Figs. 4.22 and 4.24). A *stationary* time series is one whose statistical properties such as mean, variance, and autocorrelation are constant over time. The throughput of traffic on an Internet link may be stationary if measured over short periods of time, for example an hour. However, for realistic protocol evaluations, it is useful to note that not only is Internet traffic non-stationary over longer periods, for example a day, it may even be non-stationary over shorter periods of an hour, as is the case in the hour long IBM traffic we use as input in half our experiments. This creates

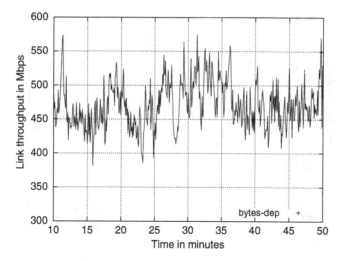

Fig. 4.21 Link throughput in Mbps – UNC (unconstrained mode)

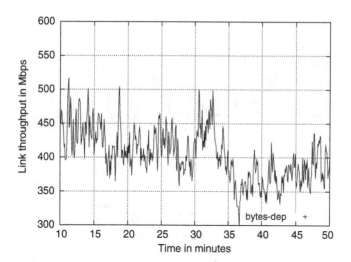

Fig. 4.22 Link throughput in Mbps – IBM (unconstrained mode)

interesting effects on the metrics in the *constrained* experiments when using the IBM traffic, especially for queue dynamics at the router.

In the following sections, we present results for two experiments: one using the UNC traffic as input, and the other using IBM traffic as input. Both experiments were run in the *unconstrained* mode. We present results for the four performance metrics. All of these results are again presented in Chapter 5. However, in that chapter, we use the control set for comparison against other models. In this chapter we present these results as a study of the control set with a focus on detailed discussion of the four performance metrics.

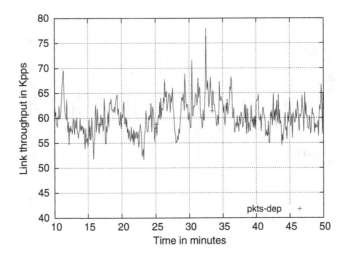

Fig. 4.23 Link throughput in Kpps – UNC (unconstrained mode)

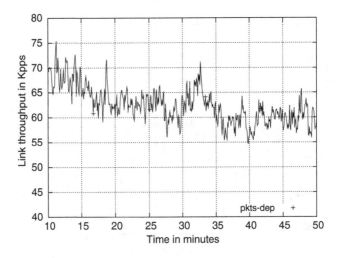

Fig. 4.24 Link throughput in Kpps – IBM (unconstrained mode)

4.5.2 Connection Duration

We define connection duration for any TCP connection as the time elapsed between the transmission of the first data byte and the receipt of the last data byte of that connection. Connection duration for every connection is measured and logged at the traffic generators. During the hour long experiment, every traffic generator creates a number of logs reporting on the performance of the TCP connections in the experiment. This includes connection duration and response times for every connection.

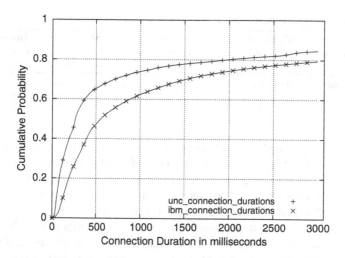

Fig. 4.25 Connection duration – CDF Control set – UNC and IBM – unconstrained

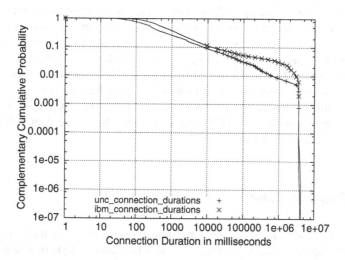

Fig. 4.26 Connection duration – CCDF Control set – UNC and IBM – unconstrained

Figures 4.25 and 4.26 compare the cumulative distribution functions (CDFs) and
the complementary cumulative distribution functions (CCDFs) for duration of the
TCP connections in the two experiments. The CDF shows a linear plot of durations
up to 3 seconds. The CCDF is on a log-log scale and shows durations up to the entire
hour of the experiment, which is 3600 seconds or 3.6 x 10^6 milliseconds. This data
shows durations for several million TCP connections – 4.7 million for the experiment using UNC traffic and 2.8 million for the experiment using IBM traffic.

As shown in Fig. 4.25, and enumerated in Table 4.1, 80 % of the connections in the
UNC replay complete in less than 2 seconds, while 80 % of the connections in the

Table 4.1 Connection duration a-t-b-t with usernet in unconstrained mode

Unconstrained experiments	Median of connection durations	80 % or less of connection durations	Mean of connection durations	Top 10 % of connection durations
using UNC traffic	260 milliseconds	2 seconds or less	33 seconds	[3] 8 seconds
using IBM traffic	550 milliseconds	3 seconds or less	87 seconds	[3] 13.5 seconds

IBM replay take 3 seconds or less to complete. The median connection durations are 260 milliseconds and 550 milliseconds for the UNC and IBM replays respectively. These distributions have long tails as shown in Fig. 4.26. Hence the average connection duration is relatively high. The average duration of the TCP connections was 33 seconds and 87 seconds for the UNC and IBM replays respectively.

A total of 10 % of the connections run longer than 8 seconds in the UNC replay and longer than 13.5 seconds in the IBM replay. There are some connections that last the entire hour of the experiment. These were connections that, as measured in the original Internet link, started at or before the start of our trace collection, and continued to transmit data up to the end of, or beyond, our hour long trace collection. Such long connections were sometimes dominated by the number of bytes transmitted; for example, a single connection transmitting a few gigabytes of data over the period of an hour. Often, however, very long duration connections, at least in the traffic we used, were dominated by long endpoint latencies with user think-times (inter-epoch latencies) of a minute or more between request-response exchanges within a connection. And as shown in Fig. 3.10 (see Chapter 3), several thousand connections in both UNC and IBM traffic had more than 100 epochs (request-response exchanges).

4.5.3 Response Time

We define response time for a request-response exchange in a sequential connection as the time elapsed between the transmission of the first data byte of a request and the receipt of the last data byte of its response. Hence, response time or epoch response time is defined only for sequential connections since concurrent connections do not have the notion of serialized request-response exchanges between the endpoints of a TCP connection. Response times are measured for every request-response exchange, and recorded in logs on the traffic generators.

It is interesting to note that while connection durations are recorded as one data point for every TCP connection in an experiment, response times are recorded as one data point for every epoch in a sequential connection. Hence, the number of response time data points in the distribution is dependent on not only the number of sequential connections but also the average number of epochs per sequential connection in the traffic being replayed. The IBM traffic had only 2.73 million sequential connections and the UNC traffic had 4.57 million sequential connections.

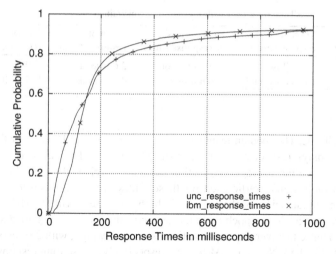

Fig. 4.27 Response Time – CDF Control set – UNC and IBM – unconstrained

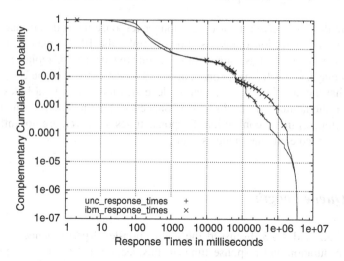

Fig. 4.28 Response Time – CCDF Control set – UNC and IBM - unconstrained

However, on average the number of epochs for the IBM connections (9 epochs per connection) is higher than that of the UNC connections (3 epochs per connection) as shown in Fig. 3.9 (in Chapter 3). Hence the UNC replay had only 13 million request-response exchanges while the IBM replay had about 24 million request-response exchanges, despite the fact that IBM traffic had only 60 % the number of connections as UNC traffic.

Figures 4.27 and 4.28 show the distributions for response times for all epochs of all sequential connections. The CDFs show response times up to 1 second. As shown in Fig. 4.27, and enumerated in Table 4.2, 80 % of the response times in the UNC replay are less than 295 ms, and 80 % of the response times in the IBM replay are

Table 4.2 Response Time for a-t-b-t with usernet in unconstrained mode

Experiment	Median of response times	80 % or less of response times	Mean of response times	Top 10 % of response times
UNC replay	110 milliseconds	295 milliseconds	2.6 seconds	[3] 800 milliseconds
IBM replay	130 milliseconds	240 milliseconds	4.4 seconds	[3] 550 milliseconds

less than 240 ms. The median response times are 110 ms and 130 ms for the UNC and IBM replays respectively. These distributions have long tails as shown in Fig. 4.28. Hence the average connection duration is relatively high. In fact, the analysis of the original traffic used for these replays revealed much longer intra-epoch endpoint latencies for the top 1 % in both traffic sets, with the IBM connections having longer intra-epoch endpoint latencies than the UNC connections (see Fig. 3.18). Hence the average response time is relatively high, with 2.6 seconds and 4.4 seconds for the UNC and IBM replays respectively. These long response times possibly indicate long server processing times from slow servers from the original connections.

We note that for a given input traffic, longer response times do not necessarily lead to longer connection durations. For example, the IBM replay had shorter response times for 80 % of its connections as compared to the UNC replay. However, the IBM replay had much longer connection durations than those of the UNC replay. We note that the duration of a connection depends on the size (total bytes) of the connection, the number of epochs in the connection and the length of inter-epoch endpoint latencies in the connection. Response times, however, are not influenced by the inter-epoch endpoint latencies at all.

4.5.4 Queue Length

Sections 4.5.2 and 4.5.3 discussed the application-level performance metrics of connection duration and response time. In this section and the next, we present results for network-level performance measures of queue length at the core router and the number of active connections in the network. During each experiment, we sampled the outbound queue at the first router (see Fig. 4.2) every 10 ms. Figure 4.29 shows the distributions for router queue lengths for both UNC and IBM replays.

Since the average throughput was 471 Mbps for the UNC replay, and 404 Mbps for the IBM replay, there was almost no congestion on this 1Gbps link. The traffic was bursty, however, and there were a few brief intervals when the network experienced spikes that were well over 1Gbps. Hence although the queue was almost always empty (about 99 % of the time), these momentary spikes led to packets being queued with roughly 10 or more packets in the queue for 0.05 % of the time for both experiments.

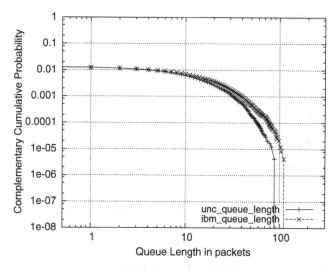

Fig. 4.29 CCDF of queue length (control set – UNC and IBM – unconstrained)

4.5.5   Active Connections

In this study, we define a connection as an 'active connection' in the network at a given time t, if the SYN for that TCP connection has been seen on the network, but the FIN or RST has not yet been recorded. Hence, an active connection could be actively sending packets or just experiencing end system or network latencies at the time that it is considered an active connection in the network.

The number of active connections in the network is directly proportional to two characteristics of the original traffic. First is the total number of connections being replayed in the hour-long experiment. Second, and more influential, is the duration of these connections. Figure 4.30 shows the time series of active connections in the two experiments. The UNC replay recorded on average 45,000 active TCP connections in the network while the IBM replay recorded on average between 68,000 and 78,800 active connections during the middle 40 minutes of the experiment.

Note the change in active connections around t = 30 minutes for the IBM replay is consistent with the non-stationarity of that traffic. The IBM traffic had fewer total connections than the UNC traffic over the hour. So, how come the IBM replay shows more active connections? Indeed, the UNC traffic consisted of 4.7 million TCP connections, while the IBM traffic consisted of 2.8 million TCP connections. However, on average, the TCP connections in the IBM traffic were longer in duration. Hence, we observe that the number of active connections in the IBM replay is much higher than that of the UNC replay.

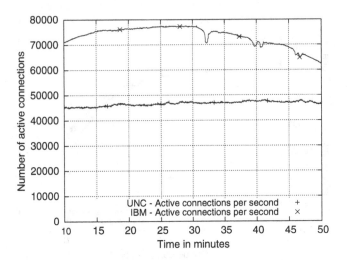

Fig. 4.30 Active connections (control set – UNC and IBM – unconstrained)

4.6 a-t-b-t with usernet in Constrained Mode

So far, we have discussed results for the replay experiments using the UNC and IBM traffic in the *unconstrained* mode, using the *a-t-b-t* connection structure model with *usernet* RTT emulation. In this section, we present the results for this control set in the *constrained* mode. For experiments in the *constrained* mode, the link bandwidth between the core routers was set to 105 % of the expected average offered load. Hence, for replays in the *constrained* mode, we set this router-to-router link to 496Mbps for UNC replay and 424Mbps for IBM replay.

4.6.1 Throughput

Figures 4.31 and 4.32 show the byte throughput time-series for the UNC and IBM replay experiments respectively. Figures 4.33 and 4.34 show the corresponding packet throughput time-series. These figures show throughput data aggregated over 5 second intervals. We show the throughput as measured in the middle 40 minutes of the experiments at the bottleneck link between the routers.

Figure 4.31 shows that the mean throughput for the UNC replay – 485 Mbps with a standard deviation of 18 Mbps. Figure 4.32 shows that the mean throughput for the IBM replay – 421 Mbps with a standard deviation of 9 Mbps. The corresponding packet throughputs are shown in Figs. 4.33 and 4.34. The mean packet throughput for the UNC replay was 61 Kpps with a standard deviation of 1.8 Kpps. And the mean packet throughput for the IBM replay was 64 Kpps with a standard deviation of 1.8 Kpps. See Table 4.3 for these throughput values.

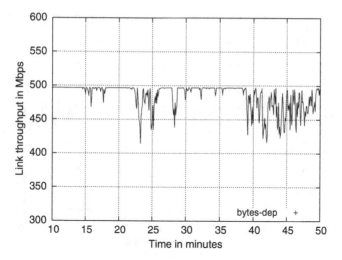

Fig. 4.31 Link throughput in Mbps – UNC Control set – UNC and IBM – constrained mode

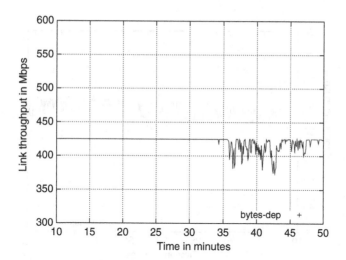

Fig. 4.32 Link throughput in Mbps – IBM Control set – UNC and IBM – constrained mode

Figures 4.31 and 4.32 demonstrate the effect of using average throughput when setting the constraints on the link bandwidth. For the IBM replay in the *unconstrained* mode (see Fig. 4.22), we noted that the mean of the throughput drops around $t = 32$ minutes. This was due to non-stationarity of the throughput time-series for the original IBM traffic. As a result, however, we note that for the first 36 minutes of the experiment, the bottleneck link is constantly utilized. This indicates the outbound queue at the router before this link rarely drained during this time. We see the direct consequence of this on the router queue length measurements shown in Section 4.6.4.

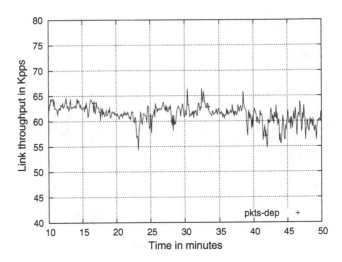

Fig. 4.33 Link throughput in Kpps – UNC Control set – UNC and IBM – constrained mode

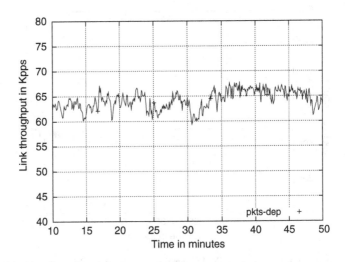

Fig. 4.34 Link throughput in Kpps – IBM Control set – UNC and IBM – constrained mode

Table 4.3 Throughput for *constrained* experiments using the control set

Constrained experiments	Mean throughput in Mbps	Standard deviation of throughput in Mbps	Mean throughput in Kpps	Standard deviation of throughput in Kpps
UNC replay	485 Mbps	18 Mbps	61 Kpps	1.8 Kpps
IBM replay	421 Mbps	9 Mbps	64 Kpps	1.8 Kpps

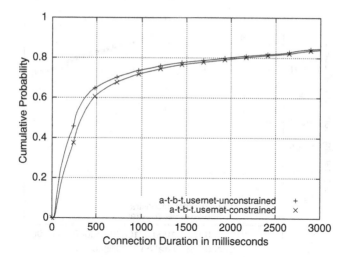

Fig. 4.35 Connection durations – UNC Control set – UNC and IBM – constrained mode

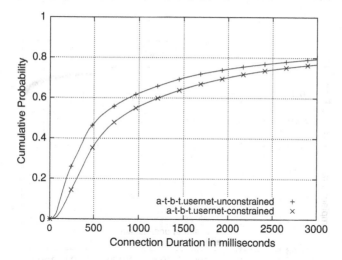

Fig. 4.36 Connection durations – IBM Control set – UNC and IBM – constrained mode

4.6.2   *Connection Durations*

Figures 4.35 through 4.38 show the distributions for connection durations for the UNC and IBM replay experiments in the *constrained* mode. The CDF shows a linear plot of duration up to 3 seconds. The CCDF is on a log-log scale and shows duration up to the entire hour of the experiment. For comparison, we have included the results for connection duration from the replays in the *unconstrained* mode.

For the UNC replay, Fig. 4.35 shows that 80 % of the connections completed in less than 2.1 seconds in the *constrained* mode compared to 2 seconds in the

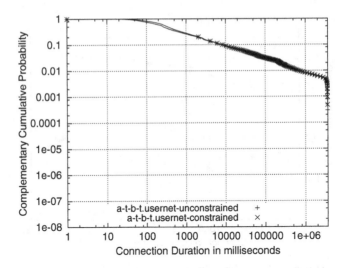

Fig. 4.37 Connection durations – UNC Control set – UNC and IBM – constrained mode

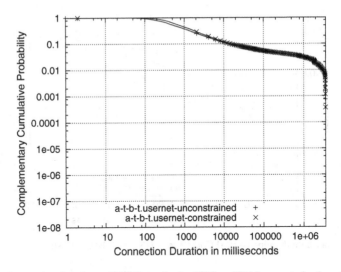

Fig. 4.38 Connection durations – IBM Control set – UNC and IBM – constrained mode

unconstrained case. For the IBM replay, Fig. 4.36 shows that 80 % of the connections took 3.9 seconds or less to complete in the *constrained* mode compared with 3 seconds in the *unconstrained* case. Clearly congestion had a slightly more debilitating effect on the IBM replay traffic than the UNC replay, though both were run with the *constrained* link set to 105 % of the average offered load. To some extent this is due to the much higher load in the experiment using the IBM trace in the first half of the experiment as compared with the second half, causing longer queuing delays in the IBM replay than in the UNC replay.

Table 4.4 Connection Duration for *constrained* experiments using the control set

Experiments	Median of connection durations	80 % or less of connection durations	Mean of connection durations	Top 10 % of connection durations
Unconstrained - UNC replay	260 milliseconds	≤ 2 seconds or less	33 seconds	≥ 8 seconds
constrained - UNC replay	330 milliseconds	≤ 2.1 seconds	33 seconds	≥ 8.3 seconds
Unconstrained - IBM replay	550 milliseconds	≤ 3 seconds	87 seconds	≥ 13.5 seconds
constrained - IBM replay	790 milliseconds	≤ 3.9 seconds	88 seconds	≥ 14.7 seconds

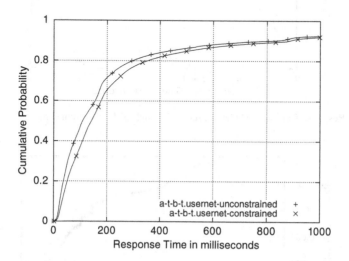

Fig. 4.39 Response Times – UNC Control set – UNC and IBM – constrained mode

   In the *constrained* mode (shown in Figs. 4.35 through 4.38 and in Table 4.4), the median connection durations were 330 milliseconds and 790 milliseconds for the UNC and IBM replays respectively. In the *unconstrained* mode, these measures were 260 milliseconds and 550 milliseconds for the two experiments respectively. Figures 4.37 and 4.38 show the long tails of these distributions. These long tails lead to high average connection durations of 33 seconds and 88 seconds for the UNC and IBM replay experiments respectively. Fully 10 % of the connections take longer than 8.3 seconds in the UNC replay and longer than 14.7 seconds in the IBM replay. In the *unconstrained* modes, these measures were 8 seconds and 13.5 seconds respectively.

4.6.3   Response Times

Figures 4.39 through 4.42 show the distributions for the epoch response times in the sequential TCP connections in the two experiments in *constrained* mode. The CDFs show response times up to 1 second. Again, for comparison we include the response time results for the experiments in the *unconstrained* modes.

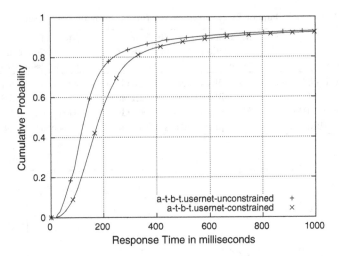

Fig. 4.40 Response Times – IBM Control set – UNC and IBM – constrained mode

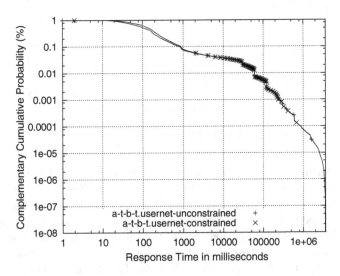

Fig. 4.41 Response Times – UNC Control set – UNC and IBM – constrained mode

As shown in these figures, and enumerated in the Table 4.5, the response times for the bottom 80 % of the response times are up 19 % and 33 % for the *constrained* experiments for the UNC and IBM replays as compared with the *unconstrained* modes for the same experiments. Clearly constraint on the link has a greater effect on response times (the time between a request-response exchange) than on connection durations. This is because connection duration is often dominated by the connection structure itself which includes the inter-epoch endpoint latencies between consecutive request-response exchanges.

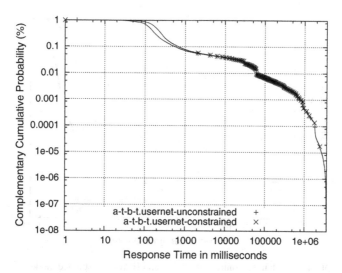

Fig. 4.42 Response Times – IBM Control set – UNC and IBM – constrained mode

Table 4.5 Response Time for *constrained* experiments using the control set

Experiments	Median of response times	80 % or less of response times	Mean of response times	Top 10 % of response times
Unconstrained – UNC replay	110 milliseconds	≤ 295 milliseconds	2.6 seconds	≥ 800 milliseconds
constrained - UNC replay	140 milliseconds	≤ 350 milliseconds	2.6 seconds	≥ 880 milliseconds
Unconstrained - IBM replay	130 milliseconds	≤ 240 milliseconds	4.4 seconds	≥ 550 milliseconds
constrained - IBM replay	187 milliseconds	≤ 320 milliseconds	4.5 seconds	≥ 660 milliseconds

The median response times for the *constrained* experiments using the UNC and IBM traffic were up 23 % and 36 % respectively from the *unconstrained* case. This is also a direct effect of the queuing delay in the network with queuing delay affecting the response time in the IBM replay more than in the UNC replay. The tails of these distributions are long but these are dominated more by the size of the data transfer and intra-epoch endpoint latencies than by the effect of queuing delay. Hence the average response times for the *constrained* experiments were similar to that of the *unconstrained* experiments.

The reason the tails of the response times seem unaffected is because the queuing delay, in the case of *a-t-b-t* connection structure experiments, represents a small fraction of the intra-epoch latencies measured for these connections in the original trace. Specifically, queuing delay is in tens of milliseconds while the intra-epoch latencies are hundreds of milliseconds to several seconds. For the top 10 % of the epochs, response times in the *constrained* mode represent an increase of 10 % and 20 % for results for the UNC and IBM replays respectively as compared with their *unconstrained* modes.

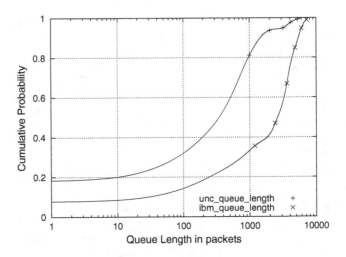

Fig. 4.43 CDF of queue length Control set – UNC and IBM – constrained mode

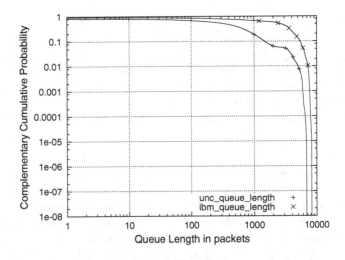

Fig. 4.44 CDF of queue length Control set – UNC and IBM – constrained mode

4.6.4   Queue Length

Figures 4.43 and 4.44 show the outbound queue at the core router before the *constrained* link. The queue was sampled every 10 milliseconds. Although both experiments were setup so that the link was *constrained* to 105 % of the average of the *unconstrained* throughput, the IBM replay saw a much longer queue. Also, the

Table 4.6 Queue length for *constrained* experiments using the control set

Constrained experiments	Queue empty / drained	Median of queue length	Mean / standard deviation of queue length	Top 10 % of queue length	Peak queue occupancy
using UNC traffic	18 % of the time	350 packets	659 / 992 packets	≥ 1460 packets	6800 packets
using IBM traffic	7 % of the time	2600 packets	2557 / 2025 packets	≥ 5400 packets	8300 packets

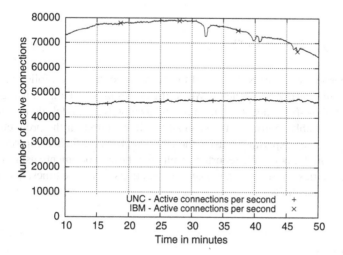

Fig. 4.45 Active connections

distribution of the inter-epoch endpoint latencies is heavier for the connections in the UNC traffic than those in the IBM traffic. This allows the queue to drain more often in the UNC replay, causing a relatively lighter queue.

As shown in Figs. 4.43 and 4.44, and enumerated in Table 4.6, the queue was empty for 18 % of the time for the UNC replay, and 7 % of the time for the IBM replay. The higher queue length for the IBM replay was partly due to the higher volume of packets and bytes in the first half of the IBM traffic. The median queue length for the UNC and IBM replays was 350 packets for the UNC replay and 2600 packets for the IBM replay.

4.6.5   Active Connections

Figure 4.45 shows the number of active connections in the network in the middle 40 minutes of the two experiments. In the *unconstrained* mode, the UNC replay recorded a median of 46,200 active TCP connections in the network, while the IBM replay recorded a median of 72,200 active connections. In the *constrained* mode, the number of active connections goes up only slightly compared to the *unconstrained* mode. This is because the queue buildup causes a small increase in the duration of connections, which leads to a small increase in the number of active connections in the network. So, in the *constrained* mode, the number of active connections had a median of 72,680 in the IBM case, but the UNC case remains roughly the same since the queuing delay was not significant enough to adversely affect the connection durations.

4.7 Chapter Summary

In this chapter, we described in detail the network configuration followed by calibration of all network components. We verified the replay of Tmix showing that the control set of *a-t-b-t* connection structure and *usernet* RTT models do indeed realistically and reliably reproduce the original traffic captured on the production link. We then presented experiments using the UNC and IBM traffic in the *unconstrained* and *constrained* modes. We reported results for these experiments using four performance metrics – throughput, connection durations, response times, queue length and active connections.

References

1. Dag: 4.3 s single channel network monitoring card. http://www.endace.com/dag-4.3s-datasheet. html. Accessed 21 June 2011.
2. Hernandez-Campos F (2006) Generation and Validation of Empirically-Derived TCP Application Workloads. Dissertation, University of North Carolina at Chapel Hill.
3. Iperf – bandwidth measurement tool. http://iperf.sourceforge.net/ Accessed 10 January 2011.

Chapter 5
Results: Effects of Round Trip Times and Connection Structures on Application and Network Performance

The principle of science, the definition, almost, is the following: the <u>test of all knowledge is experiment</u>. Experiment is the <u>sole judge</u> of scientific "truth"... Also needed is <u>imagination</u> to create from these hints [experimental results] the great generalizations – to guess at the wonderful, simple, but very strange patterns beneath them all.

– Richard Feynman [The Feynman Lectures on Physics, 1965]

In this chapter, we present results for the core set of connection replay experiments conducted for this study. We used combinations of four connection structure models, three round trip time (RTT) emulation models, two network link modes, and two sets of traffic mixes for generating traffic for these experiments. For each experiment, we studied the effect of traffic generation on four performance metrics. Two were application-level metrics: connection durations and epoch response times; two were network-level metrics: router queue length and the number of active connections in the network per unit time.

Our study leads us to two main findings. First: the RTT model used in emulating network characteristics significantly affects application and network performance. Second: the connection structure model used for generating the TCP connections affects these performance metrics even more (often orders of magnitude more) significantly than the RTT model used.

This chapter is organized as follows: in the first half of this chapter, Sections 5.1 and 5.2, we present results showing how the RTT emulation model used in traffic generation affects these four metrics in the *unconstrained* and *constrained* network modes respectively. Then, using the same set of experiments, in the second half of the chapter, Sections 5.3 and 5.4, we present the results showing how the TCP connection structure model used in traffic generation affects the same four metrics for the *unconstrained* and *constrained* network modes respectively.

J. Aikat et al., *The Effects of Traffic Structure on Application and Network Performance*, DOI 10.1007/978-1-4614-1848-1_5, © Springer Science+Business Media New York 2013

5.1 Effects of RTT Emulation Model
   in the *Unconstrained* Mode

For a given connection, we expect that the RTT will affect its duration and epoch response times. But how does using one RTT model versus another affect the aggregate distribution of connection durations and response times for a large aggregation of connections? Moreover, does the RTT model used to generate these millions of connections also affect router queue length and active connections in the network? If yes, how significant is this effect?

We quantify the answers to these questions through the results from our experiments in this and the next section, comparing the impact of three different RTT models on four performance metrics. For the first set of experiments, we assign a single RTT value for all connections using the *meanRTT* model. For the second set of experiments, we create 10 end-to-end paths in the network by emulating 10 unique delay values using the *10pathRTT* model. For the third set of experiments, we assign to each connection the specific minimum RTT found by analyzing the TCP/IP header traces using the *usernet* RTT model. For more details on these RTT models, we refer to Section 3.4 (Chapter 3).

These three RTT models create three realistic, yet significantly different, emulations of network characteristics. The *meanRTT* model emulates the network as one single path from end to end for all connections in the hour long experiment. The *10pathRTT* model is slightly more diverse and provides 10 distinct paths in the network, with discrete RTT values that are representative of measured RTTs on production links. The *usernet* RTT model is most closely representative of the original traffic being replayed. By assigning the measured RTT for each connection in the experiment, it creates a distinct end-to-end virtual path on the testbed network for each connection in the experiment. For more details on any of these (or other) RTT models, we refer to Section 3.4 (Chapter 3).

Each set of experiments in this section and the next consists of using one RTT model per experiment, keeping the TCP connection structure constant for the set. The connection structure models (described in Chapter 3) are labeled as follows in all the figures: *blk-conc* for the *block-concurrent* model which sends all bytes of a connection in both directions simultaneously, *blk-seq* for the block sequential model which sends all the bytes of a connection as one request-response exchange between the two TCP endpoints, *a-b* for the *a-b* model that emulates all epochs (request-response exchanges) from the original connection but does not model any of the endpoint latencies measured in the original connection, and finally, *a-t-b-t* for the *a-t-b-t* model that emulates all sequential epochs and concurrent ADUs as well as all endpoint latencies in every connection.

Results for experiments using the *a-t-b-t* model with *usernet* RTT, for both unconstrained and constrained modes, were presented in Chapter 4. This is the control set against which we compare all results presented in this chapter.

In the four subsections that follow, we present results for replays in the *unconstrained* mode showing the effect of using different RTT models on each of the four

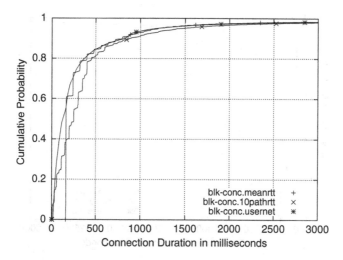

Fig. 5.1 Connection duration – UNC (*block-concurrent* connection structure)

performance measures. For all sections in this chapter we present the results for experiments using the UNC traffic as well as the IBM traffic. Unless otherwise specified, the figures on the left show results for the UNC replay, and those on the right show results for the IBM replay.

5.1.1   Effect of RTT Model on Connection Durations

In this section, we present results showing the impact of the RTT model on connection durations. We vary the RTT model per experiment while keeping the connection structure constant for that set of experiments. For example, in Fig. 5.1 we present results for connection duration for three experiments using *meanRTT* in one, *10path-RTT* in the second and *usernet* RTT in the third. All three experiments used the *block-concurrent* connection structure for generating TCP traffic.

Figures 5.1 and 5.2 show results for connection durations for varying RTT models using the *block-concurrent* connection structure for the UNC and IBM replays respectively. Similarly, Figs. 5.3 and 5.4 show results for varying RTT models using the *block-sequential* connection structure for the UNC and IBM replays respectively. Figures 5.5 and 5.6 show results for varying RTT models using the *a-b* connection structure while Figs. 5.7 and 5.8 show results for varying RTT models using the *a-t-b-t* connection structure.

For a given connection structure, we find that the RTT model impacts connection duration significantly if the duration is 500 *ms* or less. The RTT model continues to moderately impact connection durations that are up to about 1 second. But regardless of the connection structure used, the RTT model seems to have little impact on the

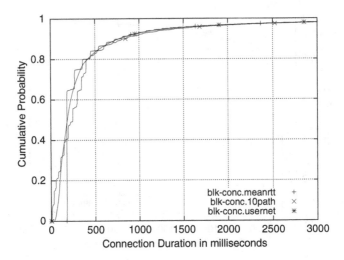

Fig. 5.2 Connection duration – IBM (*block-concurrent* connection structure)

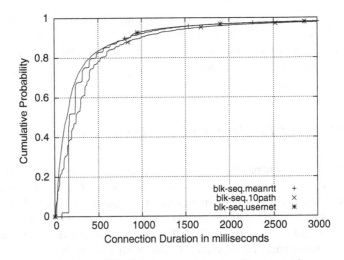

Fig. 5.3 Connection duration – UNC (*block-sequential* connection structure)

distribution for connections with duration more than 1 second. For example, for the *block-concurrent* or *block-sequential* connection structures (see Figs. 5.1 through 5.1.4), at least 98% of connections complete in 3 seconds or less with little or no difference in the distribution due to the RTT model beyond 1 second of duration.

When using the *a-b* model, as shown in Figs. 5.5 and 5.6, 97% of connections in the UNC replay complete in less than 3 seconds and 90% of connections complete in 1 second or less, regardless of what RTT model was used in the experiment. For the IBM replay experiment using the *a-b* model (Fig. 5.6), 95% of connections complete in less than 3 seconds, while only 80% of connections complete in 1 second or less.

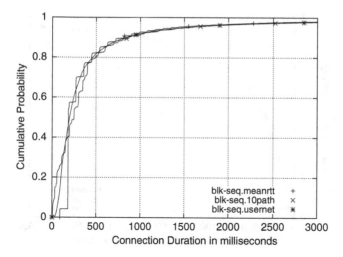

Fig. 5.4 Connection duration – IBM (*block-sequential* connection structure)

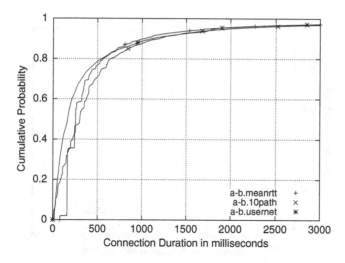

Fig. 5.5 Connection duration – UNC (*a-b* connection structure)

This difference in connection durations for the UNC versus IBM replays can be attributed to the number of epochs per connection in the two traffic mixes. 60% of connections in the original UNC traffic have only one epoch while 60% of connections in the original IBM traffic have more than one epoch. But we find that the RTT model has little impact in either set of experiments after about 1 second in the distribution of connection duration. In the replays using the *a-t-b-t* connection structure model (results shown in Figs. 5.7 and 5.8), we find that the RTT model again has a significant impact on connection durations, but only up to 500 ms and a moderate impact on durations up to 1 second.

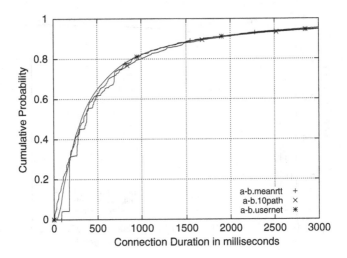

Fig. 5.6 Connection duration – IBM (*a-b* connection structure)

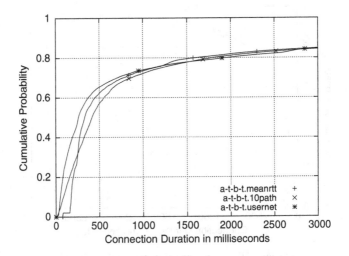

Fig. 5.7 Connection duration – UNC (*a-t-b-t* connection structure)

Since the maximum impact of RTT model is seen for connection durations up to 500 *ms*, we zoom into this part of the distribution for further discussion. See Figs. 5.9 through 5.12. These four figures show the same data as in Figs. 5.3, 5.4, 5.7 and 5.8 but we now amplify the first 500 *ms* of the distribution for connection durations; that is, we change the X-axis. We show only the *block-sequential* and *a-t-b-t* models for this discussion since the *block-concurrent* and *a-b* models have similar effects (for connection duration with RTT variation) as the *block-sequential* model.

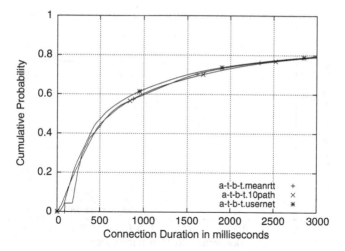

Fig. 5.8 Connection duration – IBM (*a-t-b-t* connection structure)

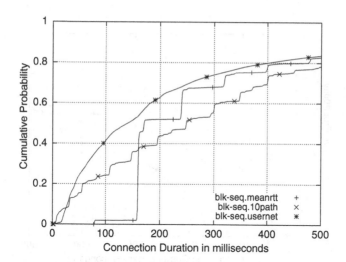

Fig. 5.9 Connection duration – UNC (*block-sequential* connection structure)

These figures show that there is a large variation among the distributions of connection duration for different RTT models used in the experiments for durations below 500 ms. Note that the mean RTT for the UNC traffic was 80 *ms* and for the IBM traffic, it was 92 *ms*. Hence, *most* connections in experiments using the *mean-RTT* model have a minimum duration of 160 *ms* (two RTTs) for the UNC replay, and a minimum duration of 184 *ms* for the IBM replay.

In a replay using the *meanRTT* model, the original connections that had a connection RTT much less than *meanRTT* now last longer and hence contribute to a

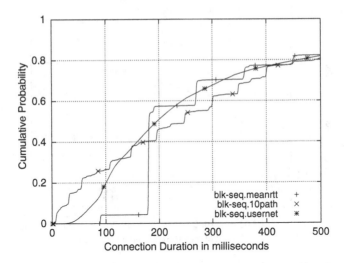

Fig. 5.10   Connection duration – IBM (*block-sequential* connection structure)

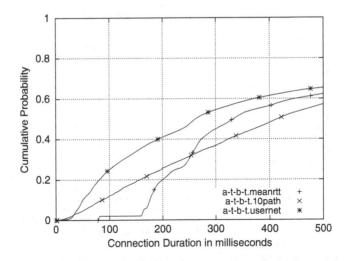

Fig. 5.11   Connection duration – UNC (*a-t-b-t* connection structure)

heavier distribution of connection duration for the initial part of the distribution.
The use of *10pathRTT* results in longer connection durations than using the *mean-RTT* or the *usernet* models. This is more so for the UNC replays than the IBM
replays. This is because the mean of the RTTs in the *10pathRTT* is 92 *ms*, which is
much higher than the mean of the RTTs (80 *ms*) for the UNC traffic. Coincidentally,
the mean of the *10pathRTT* is the same as the mean of the RTTs for the IBM traffic.
Hence, the distribution of connection durations in the IBM replays for *10pathRTT*

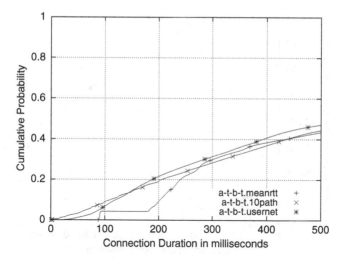

Fig. 5.12   Connection duration – IBM (*a-t-b-t* connection structure)

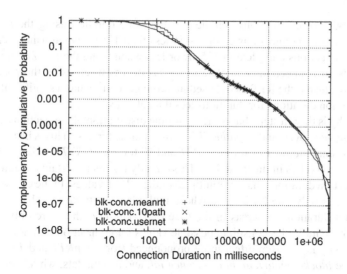

Fig. 5.13   Connection duration – UNC (*block-concurrent* connection structure)

and *usernet* RTT are closer. The *usernet* RTT shows much lighter distribution for connection duration that does the *meanRTT* model because all those connections with connection RTTs less than the mean RTT for the traffic can now replay at the rate of their original RTTs. Hence these connections complete faster with *usernet* than when using the *meanRTT* model for these same connections.

For the UNC replay with the *block- sequential* connection structure (Fig. 5.9), only 40% of connections complete in less than 160 *ms* using the *10pathRTT* model,

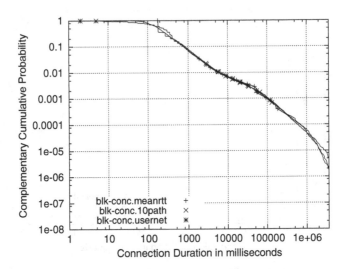

Fig. 5.14 Connection duration – IBM (*block-concurrent* connection structure)

whereas 60% of connections complete in the same duration using the *meanRTT* model. While no connections complete in less than 125 *ms* when using *meanRTT*, 50% of connections complete in 125 *ms* or less when using the *usernet* model and about 32% of connections complete in 125 *ms* or less when using the *10pathRTT* model. So, clearly, the RTT model used in traffic generation has a significant impact on connection durations for durations less than 500 ms.

But, why is there a step characteristic for the distribution of connection durations when using *meanRTT* and *10pathRTT* in most of these figures? This is because there are a very small number (1 for *meanRTT* and 10 for *10pathRTT*) of discrete values for connection RTTs in these models. This directly results in certain discrete values for connection durations that are multiples of these RTT values. In the case of *mean-RTT*, there is only one RTT value whose multiples constitute possible values for connection durations, whereas in the case of *10pathRTT*, there are only 10 RTT values whose multiples constitute possible values for connection durations.

Then, why is this step effect more pronounced (for *meanRTT* and *10pathRTT*) only in the *block-concurrent*, *block-sequential*, and *a-b* models, while barely present in the *a-t-b-t* model (see Fig. 5.11)? This is because the connections in the *a-t-b-t* model, though still dominated by their RTT for durations less than 500 *ms*, are also influenced (and more so) by the varied distribution of endpoint latencies being generated within each connection. These latencies significantly dampen the effect of a connection's RTT on its duration, thus almost eliminating the step effect for the *a-t-b-t* model. That is, due to the varied distribution of endpoint latencies which contribute to connection durations, connections emulated using the *a-t-b-t* model are not restricted to durations that are multiples of RTT alone, even when we use the *meanRTT* or the *10pathRTT* models.

Continuing discussion of Figs. 5.9 and 5.10, we observe that in the case of *usernet*, there could be as many discrete RTT values as there are TCP connections because *usernet* emulates connection RTT exactly as measured on the original network link. Hence the distribution of connection durations when using *usernet* is as diverse a set of connection duration values as the original captured traffic. However when using a small set of discrete values as in the case of *meanRTT* or *10pathRTT*, we limit the values that the distribution of connection duration can exhibit simply because connection duration can now only be some multiple of the 10 discrete values in the *10pathRTT* and the one discrete value in *meanRTT* model. This is especially true when using the block-concurrent model for the following reasons: there are no endpoint latencies, and in these experiments there is no queuing delay. Thus a connection is restricted in such cases only by how fast it can grow its congestion window to send packets. And this window growth is dependent on the connection RTT. Hence, the dominant contributor of time within a connection becomes the connection RTT. And in the absence of other time components, the duration of the connection becomes a multiple of the connection RTT. For short connections, where RTT is most dominant, this effect is seen more prominently. For connections which last longer than 1 second, the RTT model does not seem to matter. This is possibly because the size of the data transferred by the connection influences the connection duration by adding in larger amounts of transmission times relative to connection RTT. Alternately, even for small size connections, if the congestion window is relatively small, then the connection duration is increased by having to wait until acknowledgements are received before further transmission of data.

In Figs. 5.11 and 5.12, we show connection duration up to 500 *ms* using the *a-t-b-t* model with the three different RTT models of emulation. Clearly, the difference in connection duration among different RTT models is greater for the UNC replay than for the IBM replay. Again this is because the *meanRTT* value and the mean of the *10pathRTT* set of values happen to be the same for the IBM traffic. Also, as shown in Fig. 3.23 (Chapter 3), the body of the RTT distribution for UNC traffic is much lighter than that of the IBM traffic. The median connection RTT for UNC traffic was 36 *ms*, and for IBM traffic it was 68 *ms*.

There is also a much more diverse set of RTTs in the UNC traffic with a large variance in the distribution of RTTs, as compared with that of the IBM traffic. A key observation from these results is that, for a given connection structure, the distribution of connection durations and the variance in that distribution is directly related to the distribution of the connection RTTs and the variance in that original RTT distribution. Modeling RTT using the *meanRTT* or *10pathRTT* methods reduces this variance in connection RTTs and hence the resulting traffic generation produces less variance in the distribution of connection durations.

So far, we have discussed the body of the distribution of connection duration. We now study the tails of these distributions in Figs. 5.13 through 5.20. We have already found that the model of RTT emulation does not greatly affect connection durations for connections lasting more than 1 second. The tails of the distribution for connection duration shown in all these eight figures only confirm this finding.

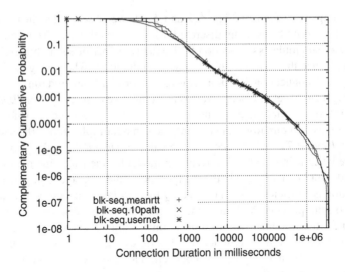

Fig. 5.15 Connection duration – UNC (*block-sequential* connection structure)

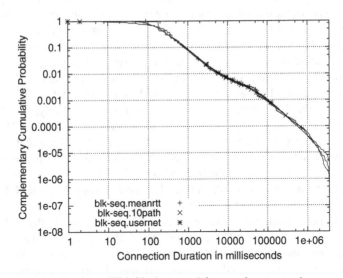

Fig. 5.16 Connection duration – IBM (*block-sequential* connection structure)

We also observe in Fig. 5.13 that for the UNC replay and the *block-concurrent* connection structure, there is a relatively quick convergence of connection durations for *meanRTT* and *usernet* beyond the initial 250 ms. This is directly because the *meanRTT* method uses the average RTT from the distribution of connection RTTs in the *usernet* model. We observe a similar convergence for these two RTT methods for the IBM replays in Fig. 5.14. Figures 5.15 through 5.20 show similar results for UNC and IBM replays using the *block-sequential*, *a-b*, and the *a-t-b-t* connection structures.

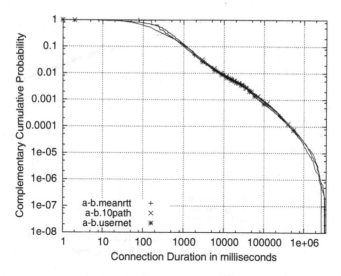

Fig. 5.17 Connection duration – UNC (*a-b* connection structure)

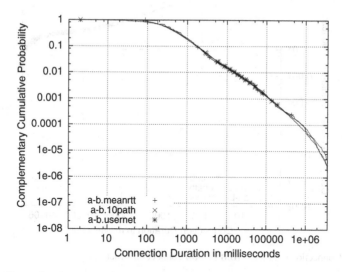

Fig. 5.18 Connection duration – IBM (*a-b* connection structure)

5.1.2   Effect of RTT Model on Response Times

In this section, we present the results of the impact of the RTT model on response times for request-response exchanges. Recall that response time is defined for each request-response exchange within a sequential TCP connection. It is the time elapsed

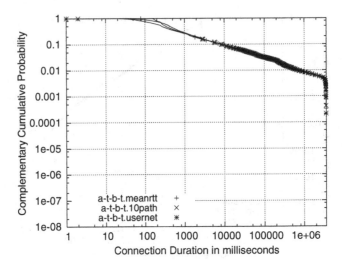

Fig. 5.19 Connection duration – UNC (*a-t-b-t* connection structure)

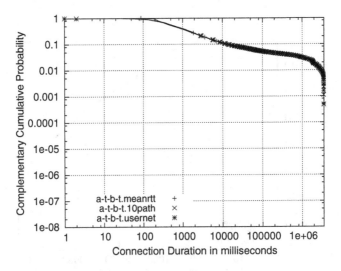

Fig. 5.20 Connection duration – IBM (*a-t-b-t* connection structure)

between the transmission of the first data byte of a request and the receipt of the last data byte of its response. Hence response time, or epoch response time, is not defined for concurrent connections or the *block-concurrent* model. For the *block-sequential* model, every connection transmits all of its data within one epoch and hence the connection duration of a connection in the *block-sequential* model is the same as its response time.

For the *a-b* and *a-t-b-t* models, there are as many response time data points in a TCP connection as there are epochs in that connection. In this section we discuss the

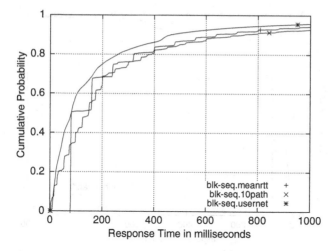

Fig. 5.21 Response Time – UNC (*block-sequential* connection structure)

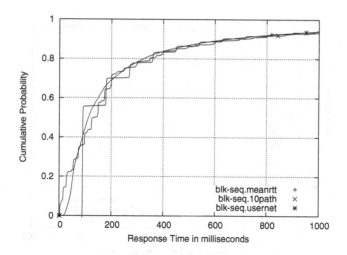

Fig. 5.22 Response Time – IBM (*block-sequential* connection structure)

impact of the RTT model on response times when using the *blk-seq*, *a-b* or *a-t-b-t*
models. Keeping connection structure the same for each set of experiments, we vary
the RTT model used for each experiment. For all the replays in this section, the data
is only for connections that were sequential in the original traffic. For example,
even for the replay using the block-sequential model, we present response time data
only for those connections that were sequential in the original traffic. This is necessary for proper comparison with other models.

Figures 5.21 through 5.26 show the distributions of response times for the UNC
and IBM replays. We observe that different RTT emulation methods clearly have

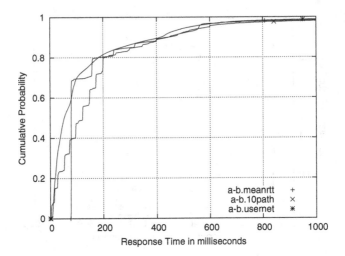

Fig. 5.23 Response Time – UNC (*a-b* connection structure)

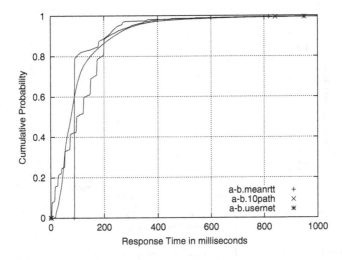

Fig. 5.24 Response Time – IBM (*a-b* connection structure)

different impact on the response times. The effect of different RTT models on response time also depends on the characteristics of the original traffic. For example, the UNC replays show greater differences in the distributions of response times due to RTT models than do the IBM replays.

For a given connection structure, we find that the RTT model impacts response time distribution significantly up to about 500 *ms* or less. As seen in Fig. 5.21, with the *block-sequential* connection structure, the RTT model continues to moderately impact response times up to about 1 second. However, for IBM replays (Fig. 5.22)

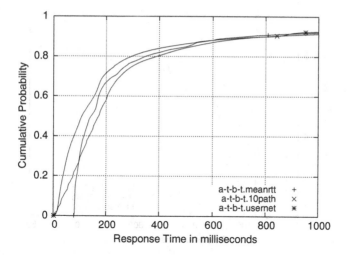

Fig. 5.25 Response Time – UNC (*a-t-b-t* connection structure)

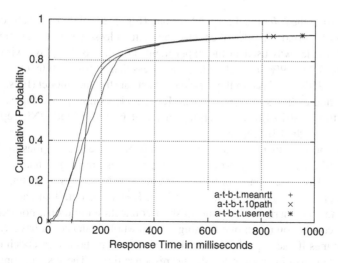

Fig. 5.26 Response Time – IBM (*a-t-b-t* connection structure)

with the block-sequential connection structure, we see that there is almost no difference among the RTT models after about 500 *ms* of response time. In Fig. 5.21, we also observe that the *usernet* RTT model causes the smallest response times followed by *meanRTT* followed by *10pathRTT*. This result is clearly because the RTT of the average connection becomes larger when using the *meanRTT* model since all the connections that would have had lesser than the mean RTT (in the original distribution) now have a greater connection RTT. Similarly, since the mean of the *10pathRTT* is the largest, the response time of the request-response exchanges using this model shows the heaviest distribution.

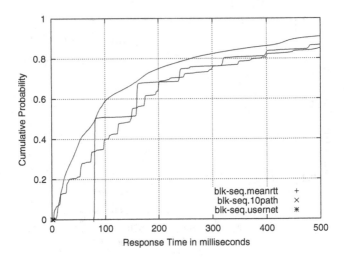

Fig. 5.27 Response Time – UNC (*block-sequential* connection structure)

When using the *a-b* model for UNC replay (Fig. 5.23), 98% of epochs complete in less than 1 second and 90% of epochs complete in less than 400 *ms*, regardless of what RTT model was used in the experiment. For the IBM replays with the *a-b* model (Fig. 5.24), 99% of epochs complete in less than 600 *ms*, while 90% of epochs complete in 250 *ms* or less. In the *a-t-b-t* connection structure model (Figs. 5.25 and 5.26), we find even lesser impact of the RTT models on response times, with the response time distributions converging at about 600 *ms* in the UNC replay, and about 300 *ms* in the IBM replay.

We also observe that, regardless of RTT model, the response times in the UNC replay using the *a-b* and *a-t-b-t* connection structures are longer than those in the IBM replay. However, connection durations in the IBM replays were longer than in the UNC ones (see figures in Section 5.1.1). Hence we note that short response times do not necessarily correspond to short connection durations. For example, a very long connection (even one running for the whole hour) could have very short response times if each epoch had small ADU sizes and short intra-epoch endpoint latencies. This would account for shorter response times. These same connections, however, could have hundreds of epochs and long inter-epoch endpoint latencies between epochs thus contributing to longer connection durations.

Clearly, the RTT emulation method has an impact on the distribution of epoch response times up to 500 *ms* or 1 second at the most. Beyond that, response times are possibly dominated by ADU sizes and intra-epoch endpoint latencies. Since the maximum impact of RTT model is seen for response times up to 500 *ms*, we zoom into this part of the distribution for further discussion below in Figs. 5.27 through 5.30. These four figures below show the same data as in Figs. 5.21, 5.22, 5.25 and 5.26 but amplify the first 500 *ms* of the distribution for response times. So, the X-axes are now up to 500 *ms* only. We show results for only the *block-sequential* and *a-t-b-t* models.

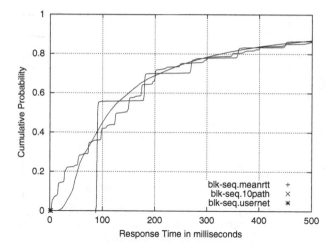

Fig. 5.28   Response Time – IBM (*block-sequential* connection structure)

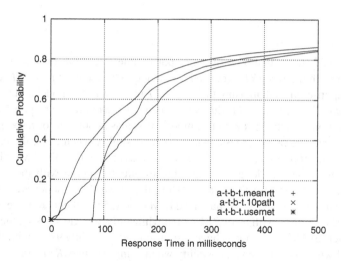

Fig. 5.29   Response Time – UNC (*a-t-b-t* connection structure)

These figures show that there is a large variation among the distributions of response times, up to 500 ms, for different RTT models used in the experiments. In Fig. 5.27, we find that RTT is a dominant time component in the request-response exchange when using the *block-sequential* model. Just as the connection durations were multiples of connection RTT for *meanRTT* and *10pathRTT* experiments, the response times are also multiples of connection RTTs for these RTT models using the *block-sequential* connection structure. This step effect is absent for response times using the *usernet* RTT model because there is a much greater variation in the distribution of connection RTTs when using the *usernet* model than when using the *meanRTT* or *10pathRTT* models.

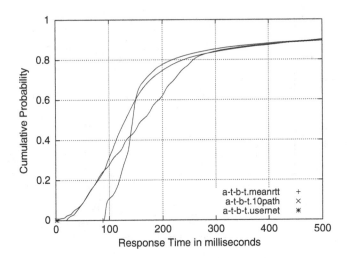

Fig. 5.30 Response Time – IBM (*a-t-b-t* connection structure)

Figure 5.27 shows that 50% of the response times are 80 *ms* or less in the UNC replay, for both the *meanRTT* and *usernet* models. The use of *10pathRTT* results in longer response times than using the *meanRTT* or the *usernet* models. The effect of using *meanRTT* over *usernet* is that 80 *ms* becomes the minimum response time for request-response exchanges with this model. Whereas about 40% of response times using *usernet* were 50 *ms* or less, that is not a possibility when using the *meanRTT* model. For the *10pathRTT* whose mean is even larger than the other two RTT models, response times are longer initially but eventually merge with the other two models. 40% of response times when using the *10pathRTT* model are 100 *ms* or less. Similarly, Fig. 5.28 shows that in the IBM replay, 55% of epochs have response times of 92 *ms* or less with *meanRTT* model while only 40% do so using the *usernet* RTT model. But 36% of epochs have response times less than 92 *ms* with the *usernet* model which is not even a possibility when using the *meanRTT* model.

The *usernet* RTT model shows the lightest distribution for response time for both the UNC and IBM replays because all the epochs with connection RTTs less than the mean RTT for the traffic now replay at the rate of their original RTTs. Hence these epochs experience faster response times than when using the *meanRTT*. Thus the distribution of response times when using *usernet* RTT is as diverse a set of possible values as the original captured traffic. For short epochs, where RTT is most dominant, this effect is seen more prominently. For epochs which last longer than 500 *ms*, the RTT model does not seem to matter as much. This is because the size of the epoch and the intra-epoch endpoint latencies (for the *a-t-b-t* model) influence the response time more than RTT does.

Figures 5.29 and 5.30 show the response times up to 500 *ms* with the three different RTT models for the *a-t-b-t* connection structure. Again, the difference in connection

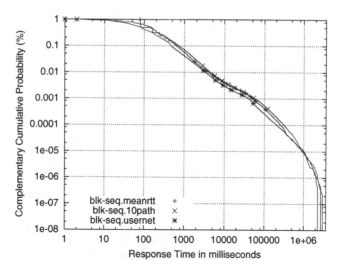

Fig. 5.31 Response Time – UNC (*block-sequential* connection structure)

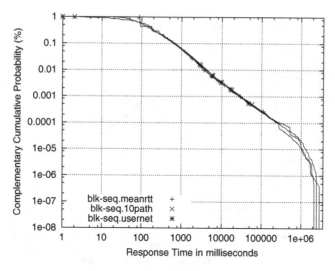

Fig. 5.32 Response Time – IBM (*block-sequential* connection structure)

duration among different RTT models is greater in the UNC replay than in the IBM replay. And the effect of RTT model on the distribution of response times diminishes after 500 ms in the UNC replays and as early as 300 ms in the IBM replays.

So far, we have discussed the body of the response time distributions. We now study the tails of these distributions shown in Figs. 5.31 through 5.36. We have already found that the model of RTT emulation does not greatly affect response time for epochs lasting more than 1 second. The tails of the distribution for response time only confirm this finding.

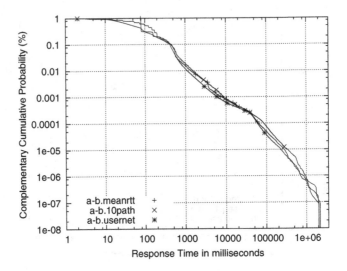

Fig. 5.33 Response Time – UNC (*a-b* connection structure)

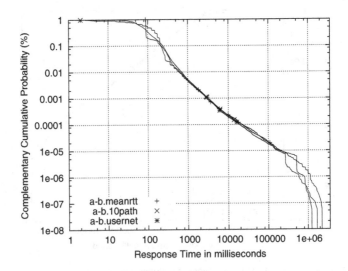

Fig. 5.34 Response Time – IBM (*a-b* connection structure)

Figures 5.31 through 5.36 show the CCDF of response times for the UNC and IBM replay experiments with the *block-sequential*, *a-b*, and *a-t-b-t* models and the three RTT emulation methods. In the *block-sequential* and *a-b* models, the RTT methods show small differences in impact on response times even for long response times. But for the *a-t-b-t* model, there is almost no difference in response time distribution. This is clearly because these long response times are dominated more by the intra-epoch endpoint latencies than the RTT of the connection.

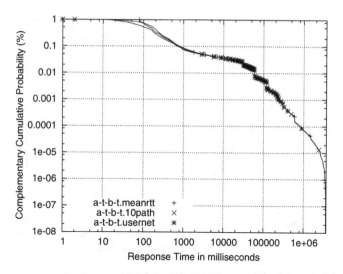

Fig. 5.35 Response Time – UNC (*a-t-b-t* connection structure)

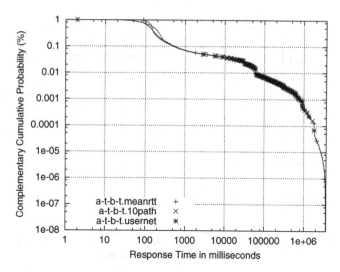

Fig. 5.36 Response Time – IBM (*a-t-b-t* connection structure)

5.1.3 Effect of RTT Model on Queue Length at the Router

In this section, we show the queue lengths at the outbound queue of the router before the *unconstrained* router-to-router link. The queue was sampled every 10 *ms* for the entire hour of the experiment. However, we only show the queue length data for the stable middle 40 minutes of the experiment. Each figure in this section shows the experimental results for a given connection structure model while varying the

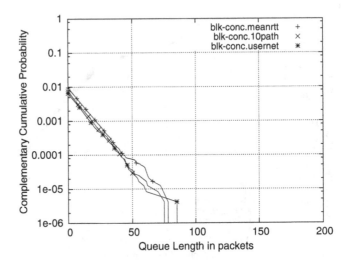

Fig. 5.37 Queue Length – UNC (*block-concurrent* connection structure)

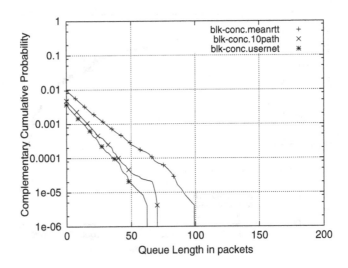

Fig. 5.38 Queue Length – IBM (*block-concurrent* connection structure)

RTT models. In Figs. 5.37 and 5.38, we show the distribution of queue length for three experiments in each set, using UNC and IBM traffic respectively. Each set of experiments used the *block-concurrent* connection structure while we varied the RTT model per experiment among *meanRTT*, *10pathRTT* and the *usernet* RTT models. Similarly, in Figs. 5.39 through 5.44, we show results for queue length for experiments varying the RTT models while keeping the connection structures constant among the *block-sequential*, *a-b*, and *a-t-b-t* models.

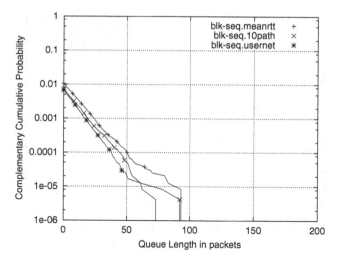

Fig. 5.39 Queue Length – UNC (*block-sequential* connection structure)

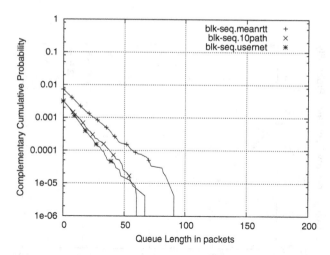

Fig. 5.40 Queue Length – IBM (*block-sequential* connection structure)

A common observation from all these experiments is that for more than 99% of the time, the queue was empty, regardless of the RTT model used for emulation. Hence Figs. 5.37 through 5.44 showing distributions of the queue length indicate almost empty queues for all those experiments. The traffic generated was bursty, however, such that even on the *unconstrained* 1Gbps link, there were momentary spikes greater than 1Gbps. Our record of the arrival pattern on the 10Gbps aggregation link before the router confirms these spikes. Hence, the tails of these distributions show a maximum queue length of around 100 packets at those momentary spikes, and 10 or more packets in the queue for about 0.05% of the time for all these replays in the *unconstrained* mode.

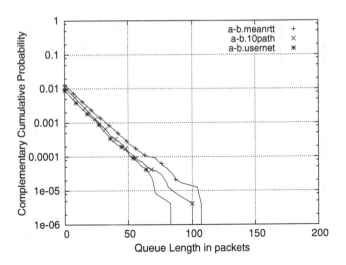

Fig. 5.41 Queue Length – UNC (*a-b* connection structure)

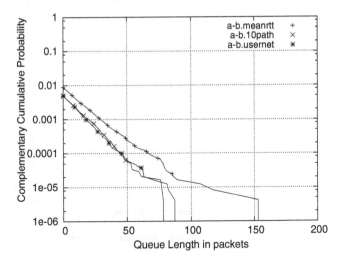

Fig. 5.42 Queue Length – IBM (*a-b* connection structure)

5.1.4   *Effect of RTT Model on Active Connections*

We define any TCP connection as an 'active connection' in the network at a given time t, if the SYN for that TCP connection has been seen on the network, but the FIN or RST has not yet been recorded. Figures 5.45 and 5.46 show the number of active connections in the network for the UNC and IBM replay experiments in the

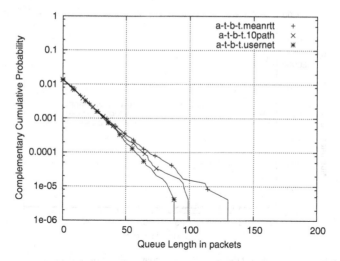

Fig. 5.43 Queue Length – UNC (*a-t-b-t* connection structure)

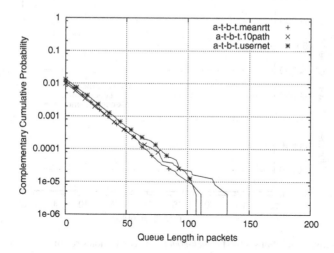

Fig. 5.44 Queue Length – IBM (*a-t-b-t* connection structure)

unconstrained mode respectively, for the middle 40 minutes of each experiment for the *block-concurrent* connection structure using a different RTT model in each of the three experiments. The RTT model clearly makes little difference in the number of active connections. Similarly, Figs. 5.47 and 5.48 show the time series of active connections for the *block-sequential* connection structure using the three RTT models. Figures 5.49 and 5.50 show the same for the *a-b* model, and Figs. 5.51 and 5.52 show the results for the *a-t-b-t* model.

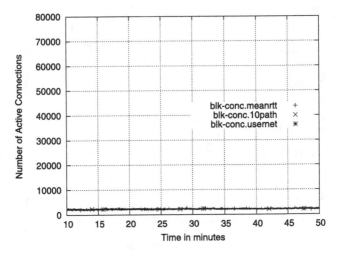

Fig. 5.45 Active connections – UNC (*block-concurrent* connection structure)

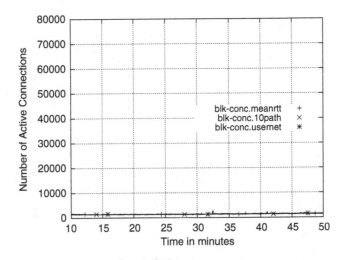

Fig. 5.46 Active connections – IBM (*block-concurrent* connection structure)

We find that, for any given connection structure, the RTT model does not affect the number of active connections in the network. This seems counter to the results that the RTT model clearly made a difference in connection durations that were 500 ms or less, and that the number of active connections in the network is directly affected by the connection durations. So, why does that difference in connection duration not manifest itself in number of active connections?

A connection is considered active during a given second whether it only lasted for 10 ms or for that whole second; so whether a connection was active for 300 ms or 550 ms on the network, it would be counted as one active connection for that second. Hence the number of active connections (measured per second as we did in

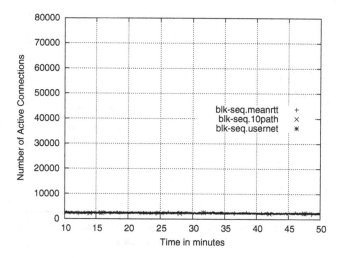

Fig. 5.47 Active connections – UNC (*block-sequential* connection structure)

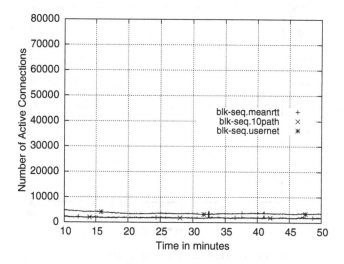

Fig. 5.48 Active connections – IBM (*block-sequential* connection structure)

this study) is a slightly gross measure of performance and is a second order effect in the network. This is why those clear differences seen in connection durations due to the different RTT models do not affect active connection counts in the network.

The number of active connections in the network is dominated by the few thousands of very long-lived connections, among the several million connections being generated over the hour, rather than the short-lived connections that replace other connections as they start and complete quickly. As an example of this effect of the long-lived connections, we observe that the number of active connections in the

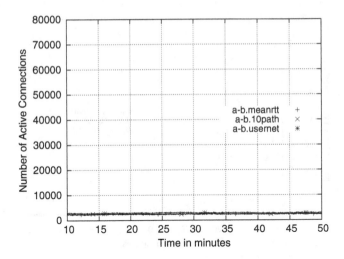

Fig. 5.49 Active connections – UNC (*a-b* connection structure)

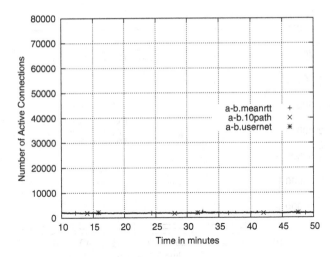

Fig. 5.50 Active connections – IBM (*a-b* connection structure)

IBM replays is almost the same as that in the UNC replays for *block-concurrent*, *block-sequential* and *a-b* models for all RTT models although the total number of connections over the hour for UNC replay was almost double that of the IBM replay. What is even more noteworthy is that the long-lived connections have such a strong impact on active connections in the network that the number of active connections for IBM replay using the *a-t-b-t* model is much higher than that for UNC replay. This is a direct consequence of the results seen in Section 5.1.1 where we found that the duration of connections using the *a-t-b-t* model was higher for IBM replay than for UNC replay.

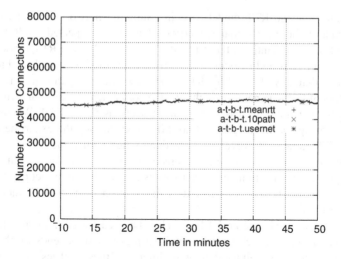

Fig. 5.51   Active connections – UNC (*a-t-b-t* connection structure)

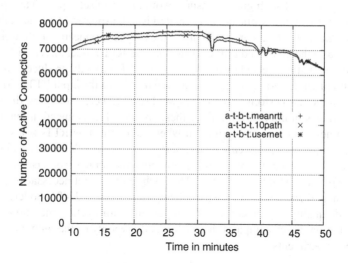

Fig. 5.52   Active connections – IBM (*a-t-b-t* connection structure)

5.1.5   *Section Summary*

In this section, we presented results for replays in the *unconstrained* mode using
UNC traffic and IBM traffic. We discussed the results for four sets of experiments
for each of the two input traffic mixes. For each set of experiments, we kept the
connection structure model the same, while varying the RTT model among the
meanRTT, the *10pathRTT*, and the *usernet* RTT models. Thus we studied the effect
of these empirically-derived RTT models on four key performance metrics: connection duration, response time, router queue length, and active connections.

We found that the RTT model used in emulating network characteristics has some impact on these performance metrics. That was an expected result. However, in this section, we quantified these results. We found that the RTT model affects connection durations and response times when these measures are less than 1 second. Beyond that, any effect of the RTT model used in an experiment is masked by other factors of traffic generation including the components of the connection structure models, which are discussed in detail in Sections 5.3 and 5.4. We also found that the router queue length showed no differences among the experiments using different RTT models. However, this was due to the fact that these were replays in the *unconstrained* mode, and hence designed to not create any queue buildup. The number of active connections in the network is a second order effect of connection durations. This metric was not affected by the differences in the RTT models used in the experiments.

So, if we had to choose an RTT model to be used for experiments, run in an *unconstrained* mode, which model would we pick? A lot depends on the performance metrics used to evaluate these experiments. If these metrics are measured at gross levels above one second, then the RTT model used may not matter. However, we would question if such gross measures would play a useful part in any protocol evaluations? If network traffic being generated is to be somewhat realistic, then it is imperative that the richness and diversity of the original connection round trip times be preserved in the generated traffic. How does the RTT model affect this?

Any metric that is affected by the connection RTT will only produce as diverse a distribution of values, for a given performance metric, as the input RTTs. For example, even for the few performance metrics we discussed here, clearly the diversity of allowable values in the distribution for these metrics, like connection durations or response times, becomes highly limited when the connection RTTs is a small discrete set of values, as was the case with *meanRTT* or *10pathRTT* models. Conversely, a rich and full set of input connection RTTs results in a similarly diverse distribution for the measured performance metric. Thus, while this is not necessarily a case of "garbage in, garbage out" since we use all empirically derived RTT models, it is still true that the quality and diversity of the inputs used for traffic generation and network emulation directly impacts the quality and diversity of the outputs measured during the experiments.

5.2   Effect of RTT Emulation Model in the *Constrained* Mode

In Section 5.1 we discussed the effect of the three different RTT emulation methods on four performance metrics: connection durations, response times, router queue length and active connections. Those were replays in the *unconstrained* mode; that is, the router-to-router link was set to 1Gbps. In this section we present results for a set of experiments run in the *constrained* mode, showing the impact of RTT models on the same four metrics; that is, the router-to-router link is set so that it is 105% of the offered load on that link. For the UNC replays in the *constrained* mode, the link

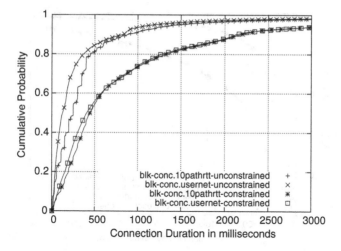

Fig. 5.53 Connection duration – UNC (*block-concurrent* connection structure)

was set to 496Mbps, and for the IBM replays in the *constrained* mode, the link was set to 424Mbps. For each set of experiments, we compare the performance metrics for different RTT models, keeping the connection structure model the same for all experiments in that set.

5.2.1   Effect of RTT Model on Connection Durations

Before we compare the effects of RTT models on connection durations for replays in the *constrained* mode, we begin by looking at the effect of such a constraint on connection duration for a given combination of connection structure and RTT. We first compare the connection durations for *10pathRTT* and *usernet* RTT models in the *unconstrained* and *constrained* modes for both UNC and IBM replays.

Figures 5.53 and 5.54 show the distributions of connection duration for four experiments each, using the UNC and IBM traffic respectively. In each figure, there are two replays in the *unconstrained* mode and two replays in the *constrained* mode. All these experiments use the *block-concurrent* model, with either the *10pathRTT* or the *usernet* RTT emulation. As observed earlier, the two experiments in the *unconstrained* mode show clear differences in connection durations between the two RTT models up to about 1 second of the distribution for connection duration. But there is a drastic shift in connection duration for both RTT emulation methods in the *constrained* mode. That is, regardless of the RTT method used, the connections experience long delays that are much greater than connection RTTs, thus causing these huge shifts in the distributions. What is causing these long delays? As we show in Section 5.2.3, the constraint on the router-to-router link results in very large queuing

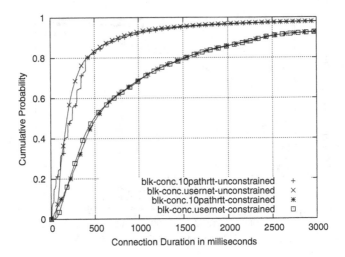

Fig. 5.54   Connection duration – IBM (*block-concurrent* connection structure)

delays in most cases. The long delays are also related to the very large queue (64 K packets) but do not have any effects from losses in TCP congestion control.

In the *unconstrained* mode for the UNC replays (Fig. 5.53), more than 80% of all connections completed in 500 *ms* or less for both the *10pathRTT* and *usernet* RTT models. But in the constrained mode, only 55% of connections completed in 500 *ms* or less using the same RTT models. There are no losses in these connections since the outbound queue at the router was set to 65,000 packets. This was done deliberately to study queuing effects due to RTT emulation. Figure 5.54 shows that the queuing delay had an even more debilitating effect on the connection durations in the IBM replays than in the UNC replay (see Fig. 5.53). In these experiments, while 82% of connections completed in 500 *ms* or less in the *unconstrained* mode, only 50% of the connections did so in the *constrained* mode. It is important to note here that although we observe the significant effect of queuing delay on the connection durations, this queue buildup and queuing delay seen by the shift in the distributions is the same for both methods of RTT emulation. As we will show in Section 5.2.3, the queue lengths and resulting queuing delays, though impacted differently by the three RTT models, are in fact a more direct consequence of the connection structure used for traffic generation. Also, the queue lengths were much greater in the case of the IBM replay experiments, partly due to the initial queue buildup since the throughput of the IBM traffic was non-stationary.

Figure 5.55 shows these queuing effects using the *block-sequential* connection structure model with the *10pathRTT* and *usernet* RTT models for the UNC replays. We observe again that the RTT emulation method has an effect on connection duration up to 500 *ms* in the *unconstrained* mode, but it has little or no effect on connection duration in the *constrained* mode. In the *unconstrained* UNC replay (Fig. 5.55), 80% of connections completed in 400 *ms* or less when using the *usernet* RTT model and 80% of connections completed in 540 *ms* or less using the

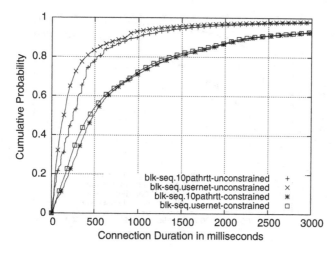

Fig. 5.55 Connection duration – UNC (*block-sequential* connection structure)

10pathRTT model. However, in the replays in the *constrained* mode, these small yet significant differences in connection durations for the two RTT models are masked by the huge effect of queuing delays on all connections in the experiment. Hence for both *usernet* RTT and *10pathRTT* models, 80% of connections complete in 1400 *ms*. Thus what was a significant 35% shift in the distribution of connection duration between the two RTT models in replays in the *unconstrained* mode is masked by the more than 85% increase in connection durations due to queuing delay in the replays in the *constrained* mode for each RTT model.

Similar effects on connection duration are seen for the UNC replay experiments using the *a-b* model with different RTT methods in *unconstrained* and *constrained* modes as shown in Fig. 5.57. In the *unconstrained* mode (Fig. 5.57), 80% of connections completed in 550 *ms* or less when using the *usernet* RTT model and in 650 *ms* or less using the *10pathRTT* model. However, in the replays in the *constrained* mode, for both *usernet* RTT and *10pathRTT* models, 80% of connections complete in 1700 *ms* or less due to the huge effect of queuing delays. The small yet significant differences in connection durations between the two RTT models are masked by the long queuing delays on all connections in the experiments.

In Figs. 5.56 and 5.58, we show results for different RTT models using the block-sequential and a-b models respectively for the IBM replays. Again we observe the huge shift in distributions for connection durations in replays in the *constrained* mode from their respective distributions in replays in the *unconstrained* mode. While this large shift is due to the large queuing delays in both sets of experiments, there is also little to no difference in the distribution of connection durations due to the RTT model being used in the replays in the *constrained* mode.

In Figs. 5.59 and 5.60 we study the results in *unconstrained* and *constrained* modes, using the *a-t-b-t* model with the *10pathRTT* and the *usernet* RTT models for

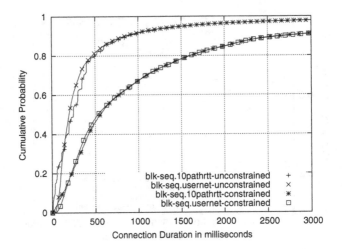

Fig. 5.56 Connection duration – IBM (*block-sequential* connection structure)

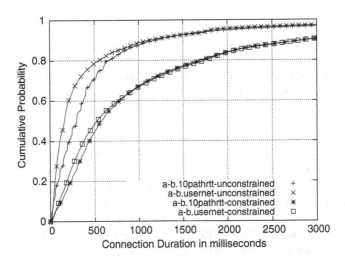

Fig. 5.57 Connection duration – UNC (*a-b* connection structure)

the UNC and IBM replays respectively. These results are significantly different from those discussed so far with other connection structures. First, there is still a small yet significant difference in connection durations between the two RTT models even in the replays in the *constrained* mode. Second, and more significant, is that the shift in the distributions between the *unconstrained* and *constrained* replays for each RTT model is much smaller than with the other connection structure models seen in Figs. 5.53 through 5.58. Why is there such a small shift? We found that this could be completely attributed to the effect of connection structure model on queuing delay. The *a-t-b-t* model creates relatively smaller queues (hence shorter

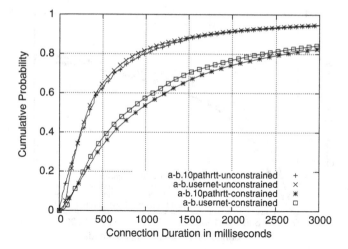

Fig. 5.58   Connection duration – IBM (*a-b* connection structure)

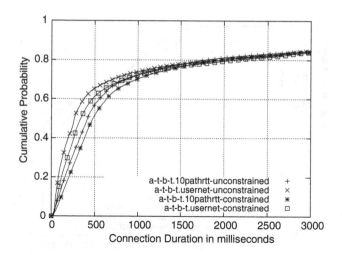

Fig. 5.59   Connection duration – UNC (*a-t-b-t* connection structure)

queuing delays) than the other connection structure models. We discuss this effect in more detail in Section 5.4.

Clearly, there is an effect from queuing on connection durations in these replays in the *constrained* mode but it is not as drastic. The difference in the connection durations for the *a-t-b-t* connection structure model using *10pathRTT* and *usernet* RTT models in the *unconstrained* modes is the same as their difference in the *constrained* modes. The effect of queuing delay is much smaller using the *a-t-b-t* model and hence the difference in connection duration between the two RTT models is preserved even in the replays in the *constrained* mode.

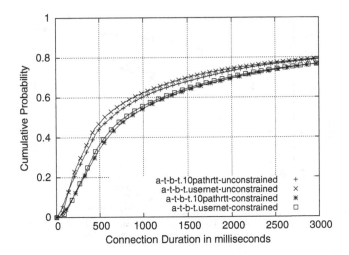

Fig. 5.60 Connection duration – IBM (*a-t-b-t* connection structure)

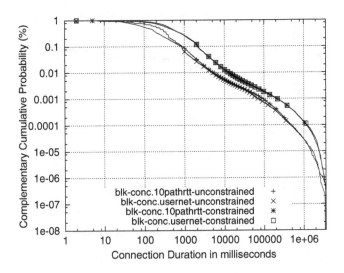

Fig. 5.61 Connection duration – UNC (*block-concurrent* connection structure)

So far, we observed the dramatic shift in the distribution of connection durations between the *unconstrained* and *constrained* replays. This effect is also seen in the tails of the distributions for connection durations as well. We show these results in Figs. 5.61 through 5.66. It must be noted that the CCDFs are in log-log scale; hence what looks like a small shift in the distributions is really a large difference in the actual distributions. And Figs. 5.67 and 5.68 show that there is not much effect of queuing delay on connections with long durations.

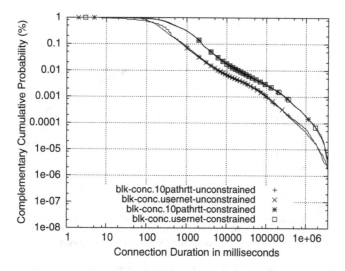

Fig. 5.62 Connection duration – IBM (*block-concurrent* connection structure)

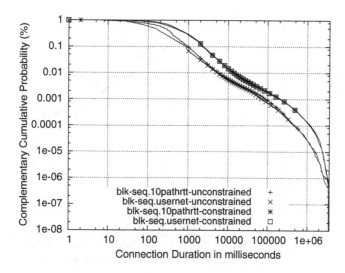

Fig. 5.63 Connection duration – UNC (*block-sequential* connection structure)

We observe that, for a given connection structure model, there is a significant effect of queuing delay in each set of experiments regardless of the RTT emulation method used in the experiment. The only set of experiments this does not hold true is the set using the *a-t-b-t* connection structure model. This is because the queuing delay and the differences in RTT emulations are insignificant latencies for these long-lived connections compared with their endpoint latencies being generated as part of the traffic model.

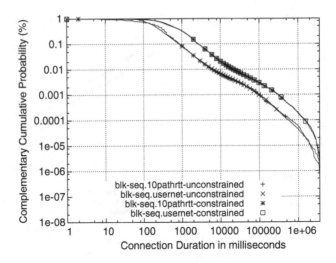

Fig. 5.64 Connection duration – IBM (*block-sequential* connection structure)

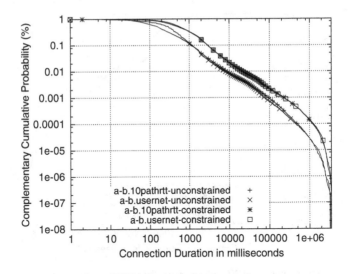

Fig. 5.65 Connection duration – UNC (*a-b* connection structure)

Having studied the queuing delays for replays in the *constrained* mode by observing the difference in connection durations between the *unconstrained* and *constrained* replays, we now discuss the direct effect of RTT emulation on connection duration for different RTT models in these replays in the *constrained* mode. Figures 5.69 through 5.76 show the distributions of the connection durations and the effects of different RTT methods for a given connection structure. For example, Fig. 5.69

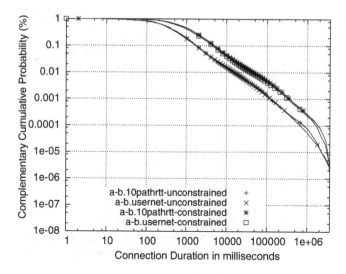

Fig. 5.66 Connection duration – IBM (*a-b* connection structure)

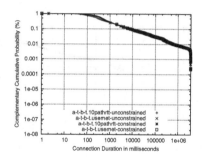

Fig. 5.67 Connection duration – UNC (*a-t-b-t* connection structure)

shows three UNC replays in the *constrained* mode, all using the *block-concurrent* structure with the *meanRTT*, *10pathRTT*, or *usernet* RTT methods of emulation.

The most remarkable observation in all these figures is that while the method of RTT emulation resulted in significantly different distribution of connection durations for durations up to 500 *ms* in the *unconstrained* modes, the RTT emulation method makes almost no difference in the replays in the *constrained* mode. The only set of replays in the *constrained* mode which show impact of RTT emulation method are the UNC replay with the *a-t-b-t* connection structure (see Fig. 5.75). And this is because the queuing delay is not significant enough in these experiments to have overshadowed the effect of differences in RTT emulation methods. The queuing delay in the IBM replay in *constrained* mode was significantly higher even with the *a-t-b-t* model and this is explained in detail in Section 5.2.3.

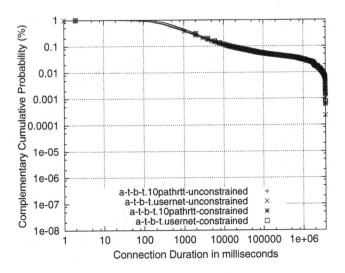

Fig. 5.68 Connection duration – IBM (*a-t-b-t* connection structure)

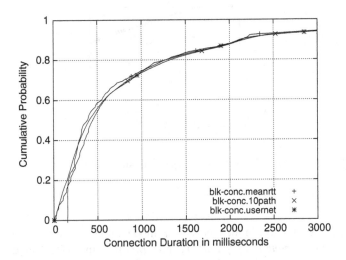

Fig. 5.69 Connection duration – UNC (*block-concurrent* connection structure)

Figures 5.77 through 5.84 show the CCDFs for connection duration for each connection structure while varying the RTT emulation method. These clearly show that the RTT emulation differences have no impact on connection duration in the tail of these distributions for replays in the *constrained* mode, for the same reasons already stated above for the body of these distributions.

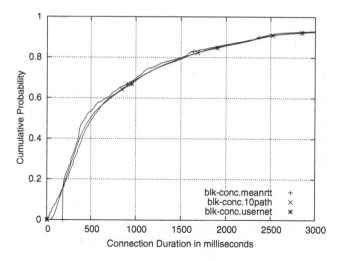

Fig. 5.70 Connection duration – IBM (*block-concurrent* connection structure)

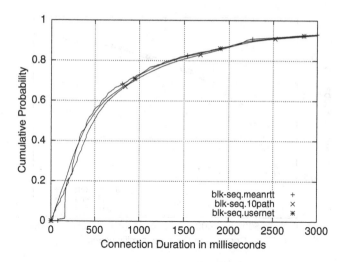

Fig. 5.71 Connection duration – UNC (*block-sequential* connection structure)

5.2.2   *Effect of RTT Model on Response Times*

In Section 5.1.2, we observed the direct effect of RTT emulation on the response times of request-response exchanges within TCP connections when there was no congestion in the network. The distribution of response time was affected by the RTT emulation method up to 500 *ms* and up to about 1 second in some cases, but was not affected beyond that. In this section, we discuss the results for a similar set

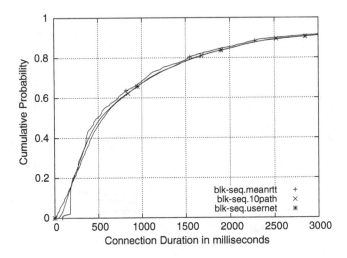

Fig. 5.72 Connection duration – IBM (*block-sequential* connection structure)

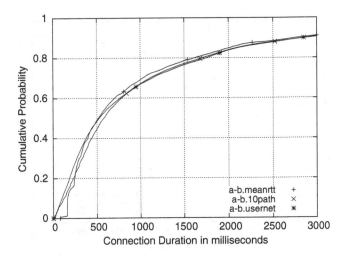

Fig. 5.73 Connection duration – UNC (*a-b* connection structure)

of experiments run in the *constrained* mode. We begin this discussion by looking at the effect of the *constrained* mode on response times. Hence, we first compare the response times for the *10pathRTT* and *usernet* RTT models for replays in the *unconstrained* and *constrained* modes.

Figure 5.85 shows the distributions of response times for four experiments using UNC traffic. There are two replays in the *unconstrained* mode and two replays in the *constrained* mode. These experiments use the *block-sequential* model with the *10pathRTT* emulation or the *usernet* RTT. The two experiments in the *unconstrained* mode show significant differences in response times between the two RTT models

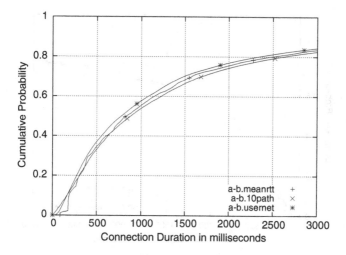

Fig. 5.74 Connection duration – IBM (*a-b* connection structure)

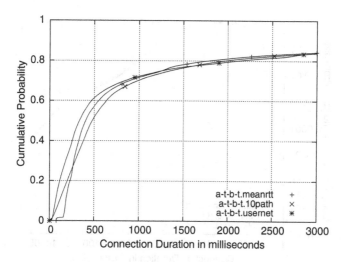

Fig. 5.75 Connection duration – UNC (*a-t-b-t* connection structure)

up to about 500 *ms* and even up to about 1 second. But there is a much larger shift in response time for both RTT emulation methods due to the congestion in the network. That is, regardless of the RTT method used, the epochs experience queuing delays that are much greater than connection RTTs, thus causing the huge shift in the distributions of response times for replays in the *constrained* mode.

In the *unconstrained* mode, shown in Fig. 5.85, roughly 80% of all request-response exchanges took about 250 *ms* using the *usernet* RTT method and about 400 *ms* using the *10pathRTT* model in the UNC replay. However, only 60% of response times are less than 400 *ms* when the same RTT methods were used in the

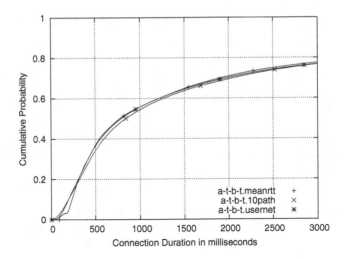

Fig. 5.76 Connection duration – IBM (*a-t-b-t* connection structure)

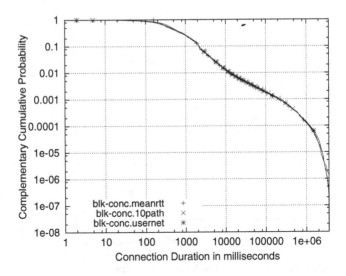

Fig. 5.77 Connection duration – UNC (*block-concurrent* connection structure)

presence of congestion in the network. 80% of these response times took up to 1 second to complete due to the effect of queuing delay in the *constrained* mode. Figure 5.86 shows that the queuing delay had an even more debilitating effect on the response times in the IBM replays with the *10pathRTT* or the *usernet* RTT. In these experiments, while 83% of request-response exchanges completed in 400 *ms* or less in the *unconstrained* mode, only 55% did so in the *constrained* mode. As we will show in Section 5.2.3, the queue lengths and hence queuing delays were much greater in the case of the IBM replays.

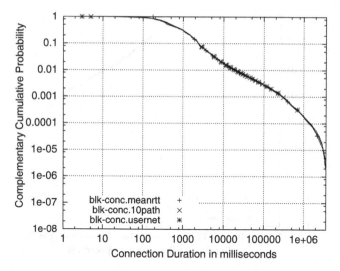

Fig. 5.78 Connection duration – IBM (*block-concurrent* connection structure)

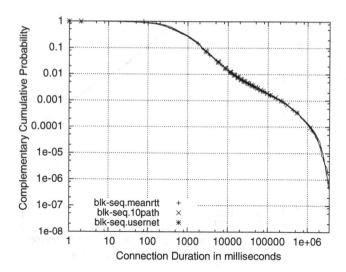

Fig. 5.79 Connection duration – UNC (*block-sequential* connection structure)

In Figs. 5.87 and 5.88 we show these queuing effects using the *a-b* connection structure model with the *10pathRTT* and *usernet* RTT models for the UNC and IBM replay experiments respectively. We observe again that the RTT emulation method has an effect on response times up to 600 *ms* in the UNC replay and 250 *ms* in the IBM replay in the *unconstrained* mode. Even in the *constrained* mode, there is clearly a difference in response times due to the two different methods of RTT

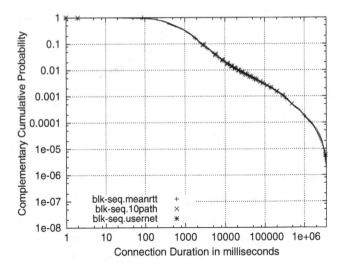

Fig. 5.80 Connection duration – IBM (*block-sequential* connection structure)

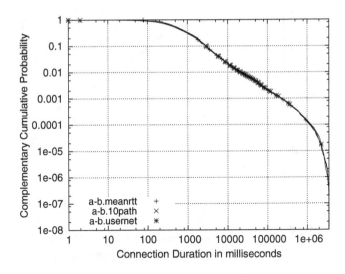

Fig. 5.81 Connection duration – UNC (*a-b* connection structure)

emulation, which is seen more clearly in the IBM case. However, the differences due to RTT emulation methods are significantly masked by the much larger effect of queuing delay on the response times in all the replays in the *constrained* mode. Thus we see that response times are not only affected by the difference in RTT emulation methods, but also have a much larger second order effect from the queuing delay.

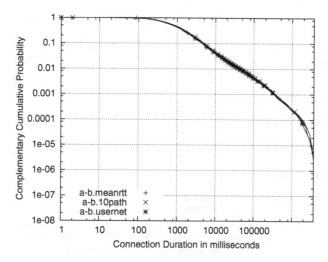

Fig. 5.82   Connection duration – IBM (*a-b* connection structure)

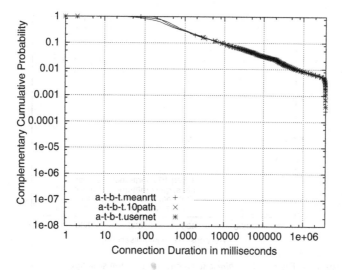

Fig. 5.83   Connection duration – UNC (*a-t-b-t* connection structure)

Figures 5.89 and 5.90 show the distributions for response times for the *a-t-b-t* model running *10pathRTT* and *usernet* RTT models in the *unconstrained* and *constrained* modes using UNC and IBM traffic respectively. While the IBM replays show the larger effect of queuing delay, there is still clearly a difference in effect on response times due to the two different RTT models even in the *constrained* mode. This is unlike the other connection structure models. As we show in Section 5.2.3, queue buildup and queuing delays are relatively smaller when using the *a-t-b-t*

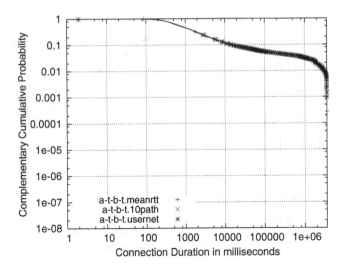

Fig. 5.84 Connection duration – IBM (*a-t-b-t* connection structure)

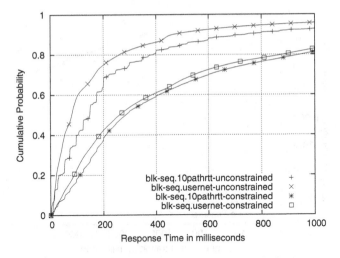

Fig. 5.85 Response Time – UNC (*block-sequential* connection structure)

connection structure model. Hence the effect on response times due to the RTT emulation methods is preserved even in these replays in the *constrained* mode.

In Figs. 5.91 through 5.96 we show the CCDFs for response times for the *10path-RTT* and *usernet* RTT models, using *block-sequential*, *a-b* and *a-t-b-t* connection structures. We observe that, for a given connection structure model, there is a significant effect of queuing delay in each set of experiments regardless of the RTT emulation method used in the experiment. The only set of experiments for which this does not hold true is the set using the *a-t-b-t* connection structure model.

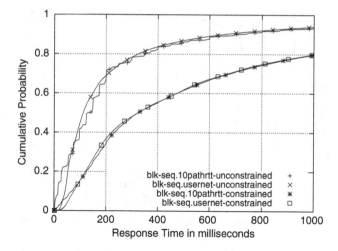

Fig. 5.86 Response Time – IBM (*block-sequential* connection structure)

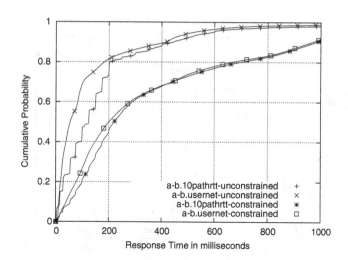

Fig. 5.87 Response Time – UNC (*a-b* connection structure)

This is because the queuing delay, as well as the delay difference among the RTT models, is insignificant for these connections compared with the endpoint latencies being generated as part of the traffic model.

Having established the dominant effect of queuing delay over RTT model on response times in replays in the *constrained* mode, we now discuss the direct effect of different RTT emulation methods on response time. Hence, we keep the connection structure model same for a set of replays. Figures 5.97 through 5.102 show the distributions of the response times as each figure shows the effect of different

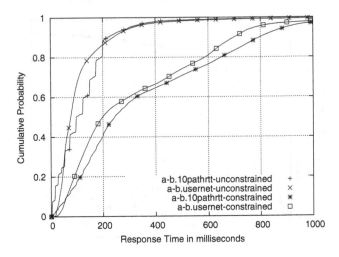

Fig. 5.88   Response Time – IBM (*a-b* connection structure)

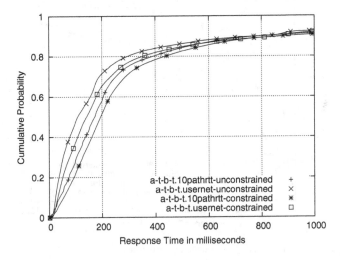

Fig. 5.89   Response Time – UNC (*a-t-b-t* connection structure)

RTT methods keeping the connection structure model same for that set of experiments.

For example, Fig. 5.97 shows three replays in the *constrained* mode, all using the *block-sequential* connection structure model while using the *meanRTT*, *10pathRTT*, or *usernet* RTT model for each experiment. We observe that the method of RTT emulation still affects, to a small degree, the response times differently up to 500 *ms*

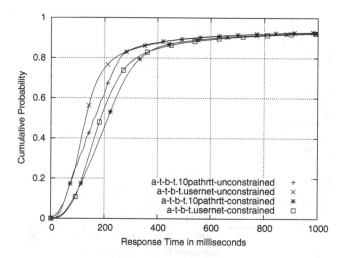

Fig. 5.90 Response Time – IBM (*a-t-b-t* connection structure)

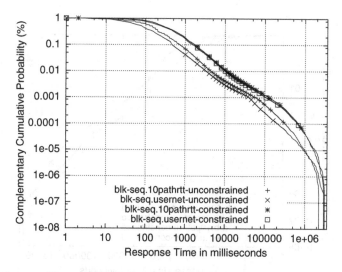

Fig. 5.91 Response Time – UNC (*block-sequential* connection structure)

even in the *constrained* mode. But there is a significant shift in overall response times due to the queuing delays.

In Figs. 5.103 through 5.108, we show the CCDFs for response time for the same set of replays in the *constrained* mode.

These clearly show that the RTT emulation differences have no impact on response times in the tail of these distributions when the experiments were run in the

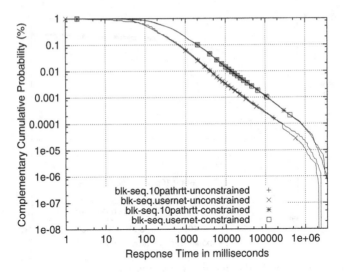

Fig. 5.92 Response Time – IBM (*block-sequential* connection structure)

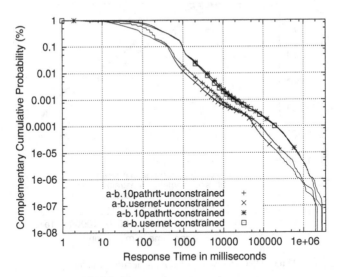

Fig. 5.93 Response Time – UNC (*a-b* connection structure)

constrained mode. As we observed in the *unconstrained* mode, the RTT model in the *constrained* mode also does not affect response times beyond 1 second in the distribution.

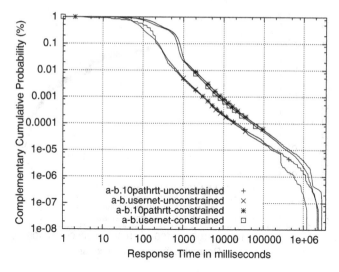

Fig. 5.94 Response Time – IBM (*a-b* connection structure)

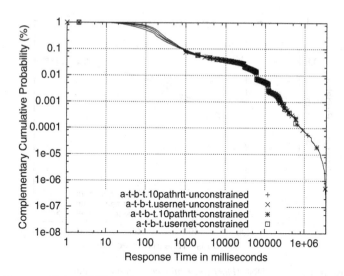

Fig. 5.95 Response Time – UNC (*a-t-b-t* connection structure)

5.2.3   *Effect of RTT Model on Queue Length at the Router*

In this section, we study the distribution of queue lengths at the outbound queue of the router before the *constrained* link. The queue was sampled every 10 *ms* for the entire hour of the experiment. However, we only show the queue length data for the

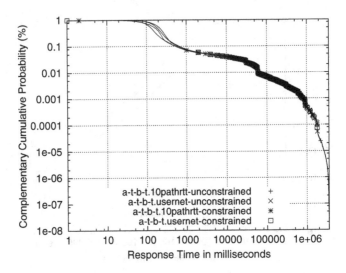

Fig. 5.96 Response Time – IBM (*a-t-b-t* connection structure)

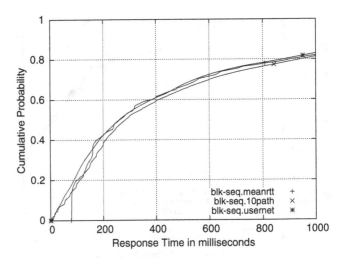

Fig. 5.97 Response Time – UNC (*block-sequential* connection structure)

stable middle 40 minutes of the experiment. Clearly, as illustrated in Figs. 5.109
through 5.116, the different RTT emulation methods have different effects on the
queue dynamics for a given connection structure model. This effect is seen for both
the UNC and IBM replay experiments. Regardless of RTT distribution the queue is
empty less than 15% of the time or less, indicating heavy queuing for all the RTT
models.

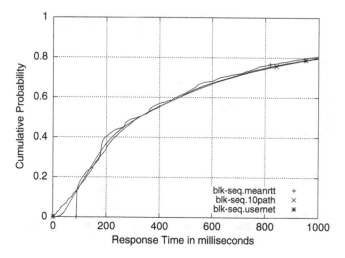

Fig. 5.98 Response Time – IBM (*block-sequential* connection structure)

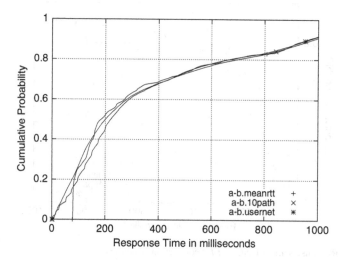

Fig. 5.99 Response Time – UNC (*a-b* connection structure)

For example, in Fig. 5.109, we show the queue lengths for the three replays in the *constrained* mode using UNC traffic, all using the *block-concurrent* connection structure while varying the RTT model used from among the *meanRTT*, *10pathRTT* and *usernet* RTT models. When using the *meanRTT* model, the router queue has less than 1000 packets for 34% of the time. However, when using the *usernet* model, the queue has less than 1000 packets only 17% of the time.

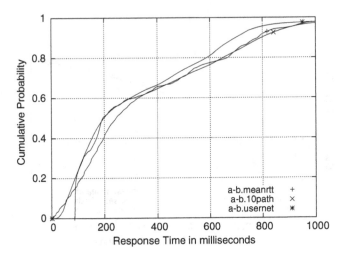

Fig. 5.100 Response Time – IBM (*a-b* connection structure)

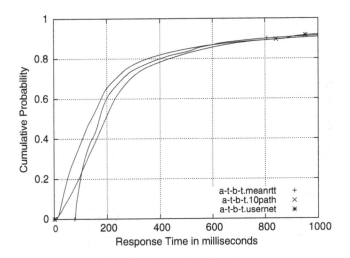

Fig. 5.101 Response Time – UNC (*a-t-b-t* connection structure)

Similarly, in Fig. 5.110 we observe that in the IBM replay with the *block-concurrent* connection structure, 33% of the time there are less than 1000 packets in the queue with the *meanRTT* model, whereas only 25% of the time, there are less than 1000 packets in the queue with the *usernet* RTT model. This clear difference in the distribution of queue length due to the different RTT models is seen for every connection structure. We show these results in Figs. 5.111 and 5.112 for the *block-sequential* connection structure, Figs. 5.113 and 5.114 for the *a-b* model and Figs. 5.115 and 5.116 for the *a-t-b-t* model.

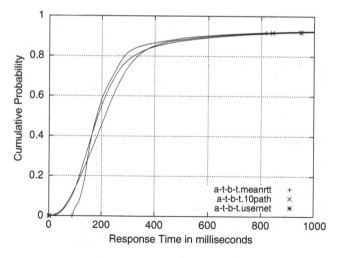

Fig. 5.102 Response Time – IBM (*a-t-b-t* connection structure)

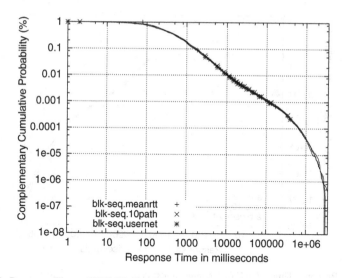

Fig. 5.103 Response Time – UNC (*block-sequential* connection structure)

There is a clear and consistent pattern in the difference in effects on queue length among the RTT emulation methods. For both UNC and IBM replays and for any given connection structure, we find the following pattern: the *meanRTT* model has the relatively lightest queue while the *usernet* model results in the relatively heaviest queue. This is because when all connections are using the *meanRTT* of 80 *ms* or 92 *ms* for connection RTTs for UNC and IBM replays respectively, there is a larger

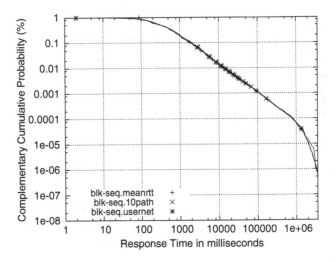

Fig. 5.104 Response Time – IBM (*block-sequential* connection structure)

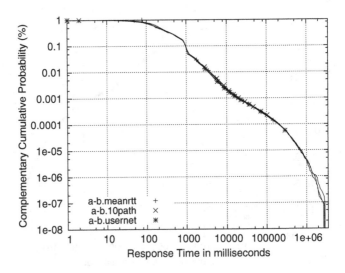

Fig. 5.105 Response Time – UNC (*a-b* connection structure)

delay between subsequent packets arriving at the router queue as compared to the other models.

For example, with the *10pathRTT* model, there are several thousands of connections with RTTs less than 80 *ms* which generate burstier arrival patterns at the router that results in more queuing. This is the same reason why the *usernet* method results in the heaviest queuing because a significant number of connections in this method

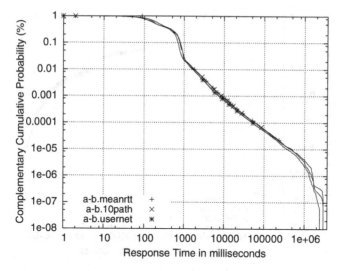

Fig. 5.106 Response Time – IBM (*a-b* connection structure)

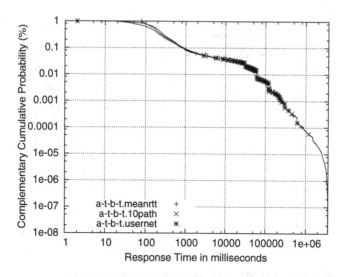

Fig. 5.107 Response Time – UNC (*a-t-b-t* connection structure)

have RTTs less than 80 *ms* (which is the mean RTT for UNC traffic), thus causing burstier arrival at the router queue. The median RTT for the *usernet* distribution is 36 *ms* for UNC and 68 *ms* for IBM. Hence half the connections in the UNC replay have RTTs less than 36 *ms*, and half the connections in the IBM replay have RTTs less than 68 *ms*, when modeling RTT with the *usernet* method.

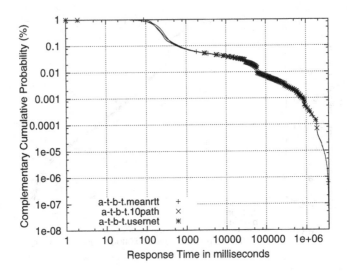

Fig. 5.108 Response Time – IBM (*a-t-b-t* connection structure)

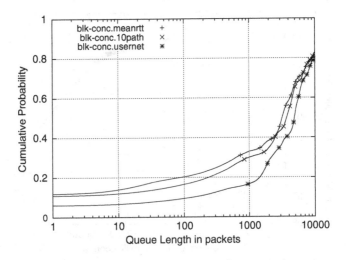

Fig. 5.109 Queue Length – UNC (*block-concurrent* connection structure)

Figures 5.117 through 5.124 show the CCDFs for the same set of experiments discussed above. Each figure shows the CCDF of queue length for three replays in the *constrained* mode using the same connection structure model but using different RTT emulation methods for the UNC and IBM replays respectively. For the *block-concurrent*, *block-sequential* and the *a-b* models, all three RTT methods result in queuing that shows over 10,000 packets in the queue for roughly 12% of the time.

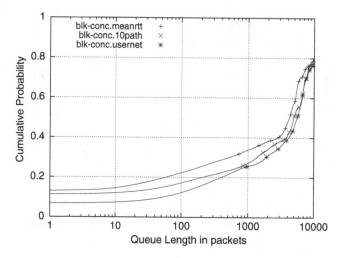

Fig. 5.110 Queue Length – IBM (*block-concurrent* connection structure)

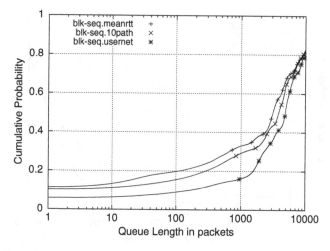

Fig. 5.111 Queue Length – UNC (*block-sequential* connection structure)

This indicates very heavy queuing due to the connection structure model, regardless of RTT model used in the experiments.

For the *a-t-b-t* model in both UNC and IBM replays, there is a small difference in the queue occupancy depending on the RTT method used. The *usernet* model accounts for the most queue occupancy since the round trip time delays between subsequent packets in a connection is more likely to be smaller in the *usernet* model than in the other two models.

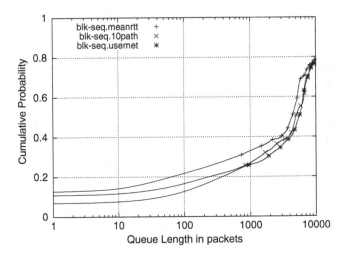

Fig. 5.112 Queue Length – IBM (*block-sequential* connection structure)

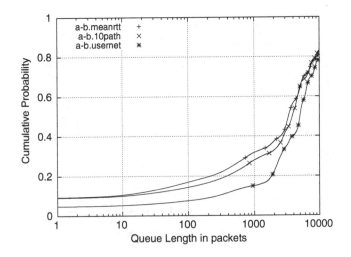

Fig. 5.113 Queue Length – UNC (*a-b* connection structure)

5.2.4   Effect of RTT Model on Active Connections

We show the number of active connections in the network for the UNC and IBM replays in *constrained* mode in Figs. 5.125 and 5.126 respectively. The data shown is only for the middle 40 minutes of each experiment for the *block-concurrent* connection structure using a different RTT method in each experiment. The RTT

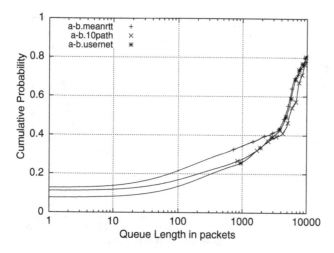

Fig. 5.114 Queue Length – IBM (*a-b* connection structure)

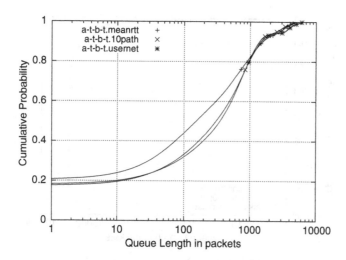

Fig. 5.115 Queue Length – UNC (*a-t-b-t* connection structure)

model clearly makes little difference in the number of active connections. Similar results are shown for the *block-sequential*, *a-b* and *a-t-b-t* connection structures in Figs. 5.125 through 5.132.

This is because connection durations directly affect the number of active connections in the network. In the replays in the *constrained* mode, we observed in Section 5.2.1 that the method of RTT emulation had less impact on connection duration for all but the *a-t-b-t* model. And in Section 5.2.3 we observed that in all the

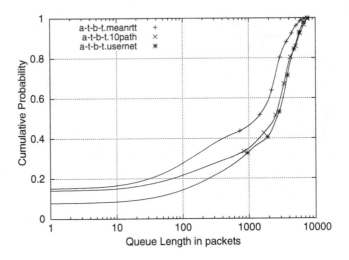

Fig. 5.116 Queue Length – IBM (*a-t-b-t* connection structure)

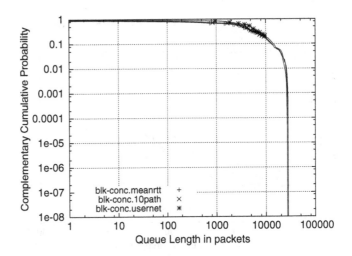

Fig. 5.117 Queue Length – UNC (*block-concurrent* connection structure)

cases other than the *a-t-b-t* experiments, there is clearly an effect of queuing for all the replays in the *constrained* mode, independent of RTT method.

That is, the number of active connections in the network is slightly higher in the initial several minutes of the experiment due to queue buildup that takes a long time to settle down. This effect is not due to the RTT emulation method but due directly to the connection structure model used. Hence we discuss this effect in greater detail in Section 5.4.4 when presenting the results of connection structure on active connections.

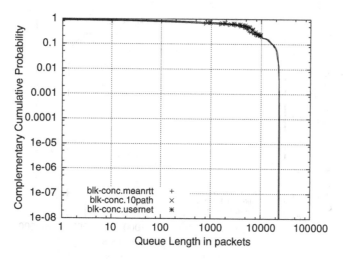

Fig. 5.118 Queue Length – IBM (*block-concurrent* connection structure)

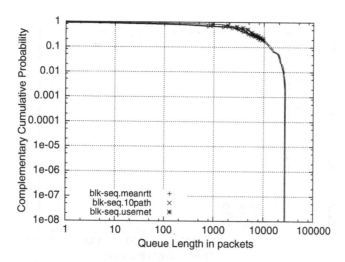

Fig. 5.119 Queue Length – UNC (*block-sequential* connection structure)

5.2.5   Section Summary

For replays in the *constrained* mode, the RTT method used for emulating network characteristics has little impact on connection durations if there is heavy congestion resulting in large queues and long queuing delays in the network. In such cases, the small effect that RTT might have had on connection durations under 1 second is mostly masked by the large effect of queuing delay on those durations.

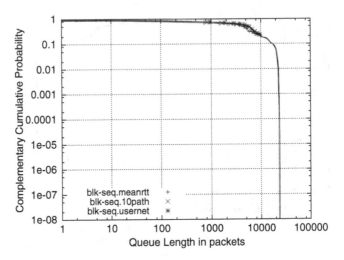

Fig. 5.120 Queue Length – IBM (*block-sequential* connection structure)

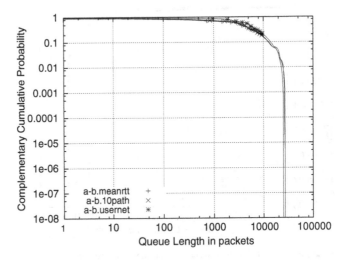

Fig. 5.121 Queue Length – UNC (*a-b* connection structure)

For response times, the RTT model has a small impact on response times less than 500 *ms* for most connection structure models, and a significant impact when using the *a-t-b-t* model. However, if there is heavy congestion resulting in large queues and long queuing delays in the network, the effect of connection RTT on response times is very small compared with the large effect of queuing delay on this metric.

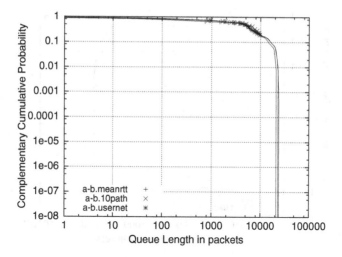

Fig. 5.122 Queue Length – IBM (*a-b* connection structure)

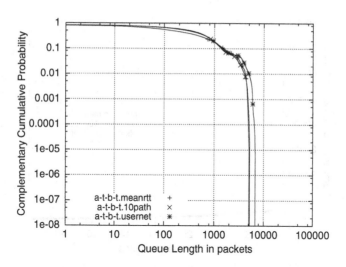

Fig. 5.123 Queue Length – UNC (*a-t-b-t* connection structure)

In this section, we also compared the effects of different RTT models on the queue length at the router. The router-to-router link was set at 105% of the offered load, and a queue capacity was 65,000 packets. For a given connection structure model, different RTT models used in generating traffic affects queue dynamics differently. In general, the *usernet* model has a diverse distribution of RTTs, and so large numbers of connections have connection RTT less than the *meanRTT* or the *10pathRTTs*. Hence we see the following patterns in all of the queue length distributions. The heaviest queue length distribution is seen when using the *usernet* model

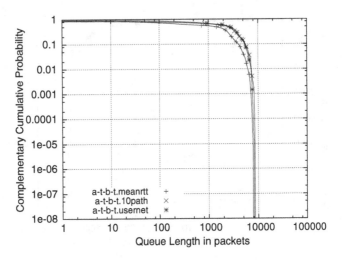

Fig. 5.124 Queue Length – IBM (*a-t-b-t* connection structure)

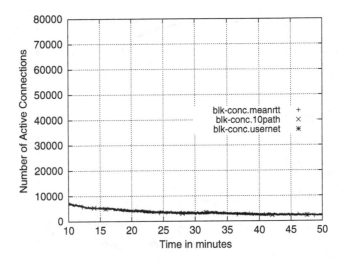

Fig. 5.125 Active connections – UNC (*block-concurrent* connection structure)

because for the large number of connections with smaller than mean RTTs, windows of packets are being sent back to back with greater frequency than with the other RTT models. That is, when the RTT is smaller, the acknowledgements from the receiver come back faster, thus allowing the sender to send another window of packets into the network. For the same reason, the next heaviest distribution of queue lengths is observed in experiments using the *10pathRTT* followed by those using the *meanRTT*. This is because with smaller RTT, a connection can grow its congestion window much faster, and thus have more packets outstanding in the network.

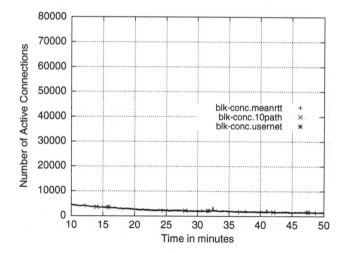

Fig. 5.126 Active connections – IBM (*block-concurrent* connection structure)

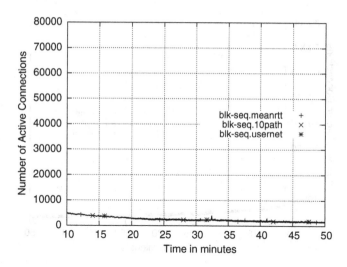

Fig. 5.127 Active connections – UNC (*block-sequential* connection structure)

In the *constrained* mode, regardless of connection structure, the queue length distributions are much heavier in the IBM replays than in the UNC replays. This is best explained by referring to the original time series of byte and packet throughput for the IBM traffic as shown in Figs. 3.1.3, 3.1.4, 3.1.7, and 3.1.8 (Chapter 3). Although the average byte throughput was measured as 404 Mbps, the time series was non-stationary. Close observations reveal that the throughput is on average higher than 404Mbps for two-thirds of the hour, and then it is on average lower than 404Mbps for the rest of the hour. For IBM replays in the *constrained* mode, the link

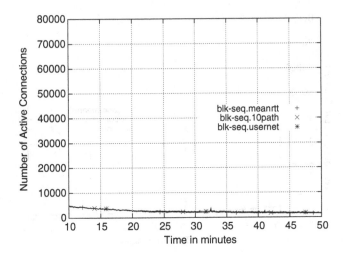

Fig. 5.128 Active connections – IBM (*block-sequential* connection structure)

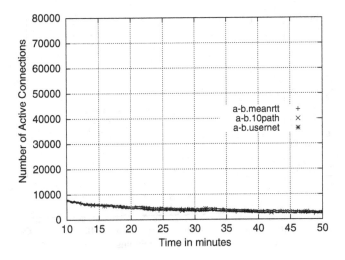

Fig. 5.129 Active connections – UNC (*a-b* connection structure)

was set to 424Mbps, so that the average offered load was effectively 95% of the router-to-router link capacity over the experiment duration. However, due to the non-stationarity, this meant that the average offered load was actually much higher than 95% for the initial two-thirds of the hour, and much lower than 95% for the last third of the hour.

Why did we then decide to use this input traffic for generating connections for this study? There were two main reasons. First, this condition did not matter for the replays in the *unconstrained* mode. Hence, we were able to use this corporate traffic

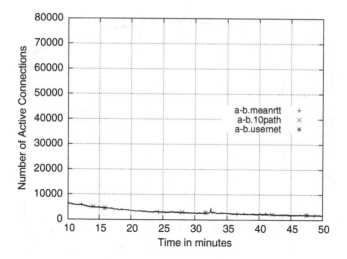

Fig. 5.130 Active connections – IBM (*a-b* connection structure)

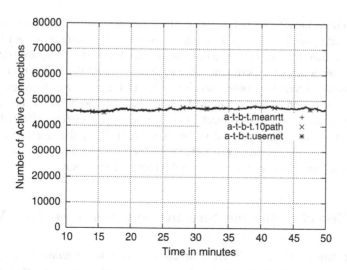

Fig. 5.131 Active connections – UNC (*a-t-b-t* connection structure)

data for this study, thus providing different traffic characteristics compared to the campus traffic data acquired at UNC. Such comparisons between different traffic inputs was helpful in understanding the outcome of experiments, and verifying the discoveries we were making about using different connection structures and round-trip times for traffic generation. Second, for replays in the *constrained* mode, this non-stationarity helped us study the effect of very severe congestion and the prolonged and debilitating effect it has on performance metrics even if the congestion is not sustained throughout the experiment.

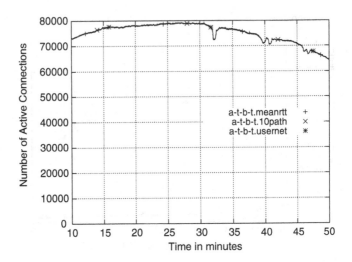

Fig. 5.132 Active connections – IBM (*a-t-b-t* connection structure)

In summary, the RTT model used in traffic generation has a significant impact on router queue dynamics. If there is a heavier distribution of connection RTTs in an experiment, that translates to more time on average between subsequent packets in a TCP connection. The experiment using such an input RTT distribution in a *constrained* mode experiences more latencies within TCP connections, thus resulting in a lighter distribution of queue lengths because there is more time for the queue to drain. The number of active connections in the network is directly affected by the duration of connections generated in the experiment. Since the effect of the RTT model on connection durations is small to none for replays in the *constrained* mode, we do not see any effect of the RTT model on active connections.

5.3 Effect of Connection Structure in the Unconstrained Mode

We experimented with four different structural models for generating a given TCP connection. As described in Chapter 3, these are the *block-concurrent* model, the *block-sequential* model, the *a-b* model, and the *a-t-b-t* model. Let us recall the basic differences among these four models. The first two connection structure models are based on only the total bytes transmitted by a TCP connection. They both transfer all bytes in both directions as one large block without internal delays in the connection. The *block-concurrent* model transfers the bytes simultaneously in both directions between the two endpoints of a TCP connection while the *block-sequential* model sends the two blocks sequentially, emulating a single request-response exchange between the two TCP endpoints.

The *a-b* model preserves the sequential exchange of bytes while further preserving the epoch structure of request-response exchanges within a TCP connection without emulating any endpoint latencies. Finally the *a-t-b-t* connection structure model not

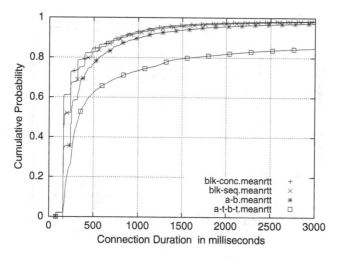

Fig. 5.133 Connection duration – UNC (*meanrtt* round trip time)

only preserves every epoch within the connection but also emulates all the endpoint latencies measured in the original traffic. These four connection structure models create four significantly different emulations for the same measured TCP connections.

So far in this chapter, we presented results for replays in the *unconstrained* and *constrained* modes to study the effects of different RTT emulation methods on the following performance measures: connection duration, response time, router queue length, and active connections. In this section and the next, we present those results for the same set of replays in the *unconstrained* and *constrained* modes, but with the goal of studying the effect of connection structure models on the four metrics mentioned above. Each set of results in this section and the next consists of presenting one RTT model per set of experiments, while varying the TCP connection structure model for each experiment. In the four subsections that follow, we present results for replays in the *unconstrained* mode showing the effect of using different TCP connection structure models on each of the four performance measures. Note that the results presented in Sections 5.3 and 5.4 are exactly the same as those in Sections 5.1 and 5.2 respectively. They are presented here with different organization of figure content to make the presentation about differences in impact of connection structures clearer.

5.3.1   Effect of Structure on Connection Durations

In this section, we show the effect of different connection structure models on the duration of connections. For example, in Fig. 5.133 we present results for connection duration for four experiments using *block-concurrent* in one, *block-sequential* in the second, *a-b* in the third, and *a-t-b-t* in the fourth experiment. All four experiments used the *meanRTT* model for emulating network characteristics.

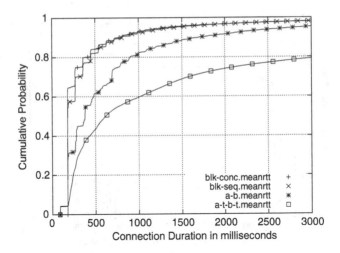

Fig. 5.134 Connection duration – IBM (*meanrtt* round trip time)

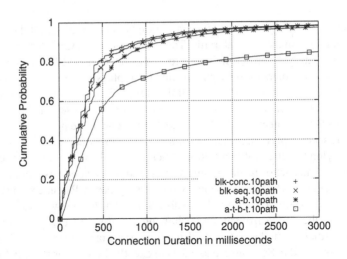

Fig. 5.135 Connection duration – UNC (*10path* round trip time)

In Figs. 5.133 and 5.134 we show results for varying connection structure models using the *meanRTT* emulation method for the UNC and IBM replays. Similarly, Figs. 5.135 and 5.136 show results for connection duration, varying connection structure models using the *10pathRTT* model in every experiment. Figures 5.137 and 5.138 show results varying connection structure models using the *usernet* RTT model.

As seen in Figs. 5.133 through 5.138, for both the UNC or the IBM replays, the *block-concurrent* and *block-sequential* connection structures result in very similar distributions of connection duration for a given input traffic and a given RTT model. The connections in the *block-concurrent* model finish slightly faster than those in

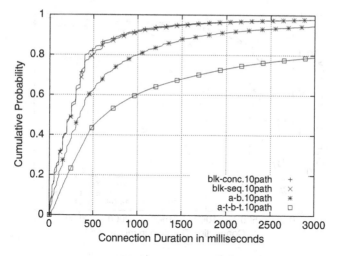

Fig. 5.136 Connection duration – IBM (*10path* round trip time)

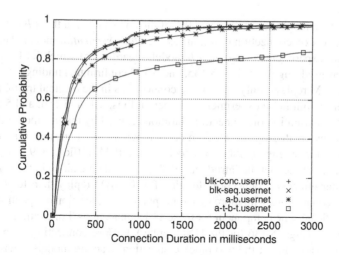

Fig. 5.137 Connection duration – UNC (*usernet* round trip time)

the *block-sequential* model because the bytes are transferred simultaneously in the *block-concurrent* case. For both these models, a little over 90% of connections complete in less than one second for both traffic inputs. This holds for all RTT models – the *meanRTT*, *10pathRTT* and *usernet* RTT models.

Studying Fig. 5.133 (UNC replay using *meanRTT*), we find that the *a-b* model takes slightly longer than the block models because the *a-b* model preserves the epoch structure of the original connection thus adding a small component of time into the connections. In the UNC replay, fully 60% of sequential connections had

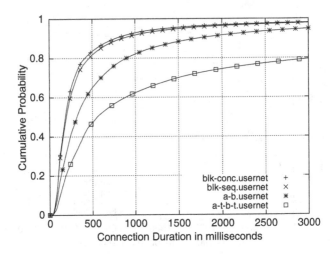

Fig. 5.138 Connection duration – IBM (*usernet* round trip time)

only one epoch and hence the fastest 60% of the connections in the *a-b* model have the same effect on connection durations as the *block-sequential* model. This result holds for all experiments using the *a-b* model, regardless of which RTT emulation method was used. Figures 5.133, 5.135, and 5.137 confirm this finding.

For the IBM replays, only 44% of the connections in the original traffic had one epoch. Hence with these experiments, as seen in Figs. 5.134, 5.136, and 5.138, we see that the distribution of connection durations caused by the *a-b* model diverges from the ones created by the block models much sooner than that for the UNC replay. Also, the average number of epochs in the IBM traffic was 9; for UNC, it was 3. Hence we see that the introduction of epoch structure alone in the *a-b* model has a greater effect on connection duration for the IBM replay than for the UNC replay. For connections with more than one epoch, the generation of epoch structure adds a significant time component for these TCP connections. For example, with the *usernet* model (Figs. 5.136 for the UNC replay), 94% of connections complete in less than 1*s* using one of the two block connection structure models whereas only 88% of connections complete in less than 1*s* using the *a-b* model. Similarly for IBM replays using *usernet* (Fig. 5.137), roughly 92% of connections complete in less than 1*s* using one of the block connection structure models but only 82% of connections complete in less than 1*s* using the *a-b* model. In the IBM replay, this larger difference is clearly due to the larger number of epochs on average in these connections and the additional time taken to replay these epochs faithfully.

For all UNC and IBM replay experiments, however, we see that the distribution of connection durations using the *a-b* model eventually lessens the gap with connection durations created by the block structure models. We conjecture that this is because not all short connections are due to small number of epochs. There are many very long connections with a small number of epochs but with a very large number of bytes to be transferred. Such large and long-lived connections would result in similar, though not same, connection durations when using the *a-b* or one of the block models.

In Figs. 5.133 through 5.134, for experiments using *meanRTT* or *10pathRTT* models, we see a *step-effect* in the distribution of connection durations. That is, there are only certain discrete possible values for connection durations with these RTT models (discussed in Section 5.1). This step effect is prominent in the block models because, in the absence of other time components within the generated TCP connections, the block models are most heavily influenced by the connection RTT for the resulting connection duration. This step effect is dampened as we add in epochs and endpoint latencies because these structural components add variance to the distribution of latencies within each TCP connection. These other latencies dampen the otherwise dominant influence of the round trip time latency for these TCP connections.

When we include the endpoint latencies (both server times and user thinktimes), we observe the most significant impact on connection duration. This effect is obvious from the experimental results using the *a-t-b-t* model. While 86% or more of connections finished in 1 second or less without endpoint latencies for all UNC replays regardless of RTT models (see Figs. 5.133, 5.135 and 5.137), only 62%, 57% and 63% of connections completed in less than a second for the replays using the *a-t-b-t* model with the *meanRTT*, *10pathRTT* and *usernet* RTT models respectively. This is the effect of including endpoint latencies in the modeling of connection structure. Thus, we find that including endpoint latencies in TCP traffic generation plays a highly significant role in the resulting distribution of connection durations. These results are even more significant for the IBM replays, where only 60% of connections completed in less than a second when endpoint latencies are included in the modeling of connection structure for all RTT emulations.

There is a slightly greater difference in the distribution of connection durations between using the *a-b* and *a-t-b-t* models with IBM traffic as compared to using UNC traffic. This is directly dependent on the slightly larger average intra-epoch endpoint latencies for IBM connections. These latencies are not modeled in the *a-b* connection structure. And the much more significant shift of the *a-b* and *a-t-b-t* models indicating longer connections for the IBM replay than for UNC replay is also due to the larger number of epochs in the connections in the IBM traffic.

So far, in this section, we have discussed the body of the distributions of connection durations. We now study the tails of these distributions. The CCDFs show results for the tails in Figs. 5.139 through 5.144. The two block models have the same impact on connection duration while connections using the *a-b* model take slightly longer to complete. Note that even the small difference seen in the distributions between the block models and the *a-b* model is significant since the axes are on log-log scale.

Here again, we observe a greater difference in impact by using the *a-b* model than by using one of the block models in the IBM replays. This is due to a high average number of epochs (9 per connection) in the IBM traffic. Addition of endpoint latencies in the *a-t-b-t* model has the largest effect for these long connections.

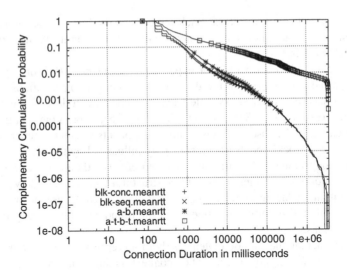

Fig. 5.139 Connection duration – UNC (*meanrtt* round trip time)

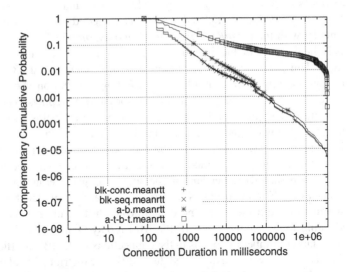

Fig. 5.140 Connection duration – IBM (*meanrtt* round trip time)

For example, in Fig. 5.139, we find that roughly 0.5% of connections take more than 10 seconds to complete when using one of the two block connection structure models while 1% of connections take longer than 10 seconds when using the *a-b* model.

With the *a-t-b-t* model, however, as much as 10% of the connections take 10 seconds or longer to complete. The top 10% of connections have durations greater than 1 second when using the *block-concurrent* and *block-sequential* models, while

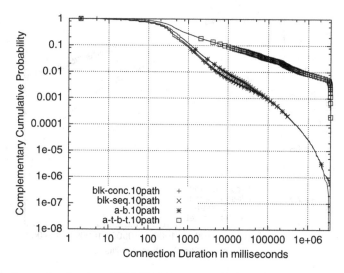

Fig. 5.141 Connection duration – UNC (*10path* round trip time)

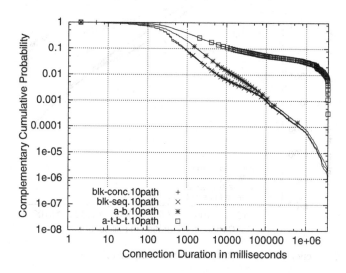

Fig. 5.142 Connection duration – IBM (*10path* round trip time)

they take about 3-5 seconds when using the *a-b* model. These connections take 10 or more seconds to complete when we add in endpoint latencies using the *a-t-b-t* model for connection structure. These results hold true for all the experiments discussed here, using either the UNC traffic or IBM traffic, and regardless of the RTT emulation method used.

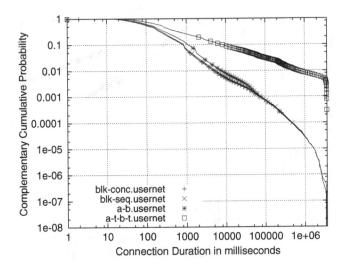

Fig. 5.143 Connection duration – UNC (*usernet* round trip time)

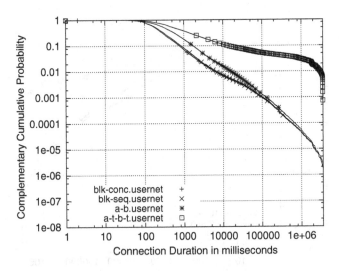

Fig. 5.144 Connection duration – IBM (*usernet* round trip time)

10 seconds for a connection to complete seems like a rather long time. The top 10% of connection durations using the *a-t-b-t* model for the UNC and IBM replays are greater than 10 seconds. What might be the contributing factors? This compares to the top 10% of the distribution of intra-epoch endpoint latencies which were greater than 1 second and 1.1 second for the original UNC and IBM traffic respectively. Similarly, the top 10% of the distribution of user thinktimes or inter-epoch latencies were 7.5 seconds and 3 seconds for the UNC and IBM traffic respectively. Concurrent connections in the UNC and IBM traffic had 14 seconds and 60 seconds respectively for the top 10% of endpoint latencies within those connections.

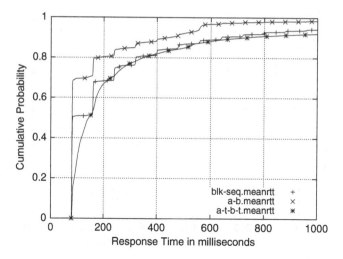

Fig. 5.145   Response Time – UNC (*meanrtt* round trip time)

5.3.2   Effect of Structure on Response Times

In this section, we present the results of the impact of the four connection structure models on the response times for request-response exchanges. Since response time is defined for each request-response exchange within a sequential TCP connection, and since the *block-concurrent* model does not generate bytes in a request-response sequence, response time is not defined for concurrent connections or the *block-concurrent* model. For the *block-sequential* model, every connection transmits data as one epoch and hence the connection duration of a connection in the *block-sequential* model is the same as its response time. For the *a-b* and *a-t-b-t* models, there are as many response time data points in a TCP connection as there are epochs in that connection.

In this section we discuss the impact of the connection structure model on response times. These figures show response times for request response exchanges for every epoch in sequential connections. Figures 5.145 through 5.150 show the distributions of response times for the UNC and IBM replays, varying connection structure models while keeping the RTT emulation method the same for each set of experiments. Overall, we observe that different connection structure models clearly have different impacts on the response times. The effect of different connection structure models on response time also depends on the characteristics of the original traffic. Hence, the UNC replays show slight differences in the impact of these models on the distributions of response times than do the IBM replays.

In Fig. 5.145, we show the results of response times for the three connection structure models, all using the *meanRTT* emulation. The *a-b* model shows much faster response times since there are no endpoint latencies within these epochs. Fully 80% of response times are 160 *ms* or less using this model. There is, however,

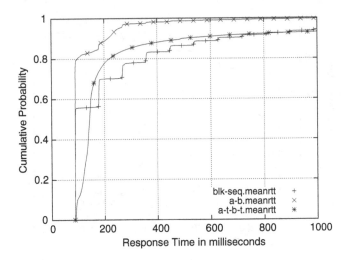

Fig. 5.146 Response Time – IBM (*meanrtt* round trip time)

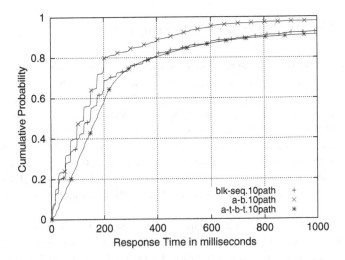

Fig. 5.147 Response Time – UNC (*10path* round trip time)

a significant difference in response times between the *a-b* and *a-t-b-t* models, which
is the consequence of modeling intra-epoch endpoint latencies in the *a-t-b-t*
model. Thus only 52% of response times are 160 *ms* or less when using the
a-t-b-t model. The *block-sequential* model does not include these latencies but has
much larger data transfers since every TCP connection in the *block-sequential*
model transfers all its bytes as a single request-response exchange; hence these
response times are longer than those using the *a-b* model.

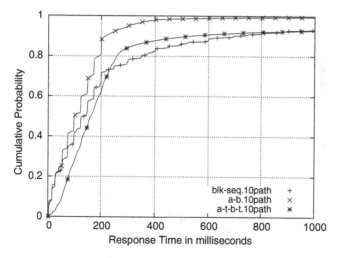

Fig. 5.148 Response Time – IBM (*10path* round trip time)

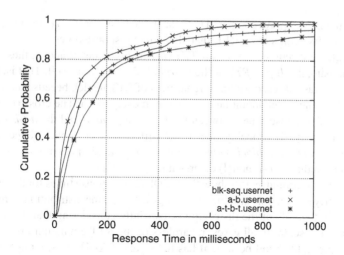

Fig. 5.149 Response Time – UNC (*usernet* round trip time)

In Fig. 5.146, we show similar results for the IBM replays. The *a-b* model again results in much shorter response times than the other two connection structure models. But what is equally significant is that the response times here are faster than in the UNC replay. This is directly due to the request sizes and response sizes in these epochs. While 80% of request sizes were less than 1000 bytes for UNC traffic, they were less than 466 bytes for IBM traffic. And while 80% of response sizes were less than 4 KB for UNC traffic, they were less than 680 bytes for IBM traffic (Figs. 3.1.11 and 3.1.13 in Chapter 3). This may explain the much faster response times for

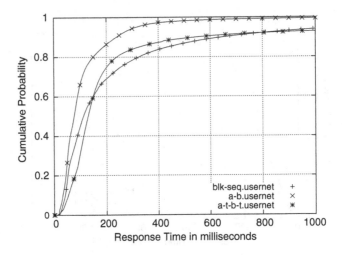

Fig. 5.150 Response Time – IBM (*usernet* round trip time)

epochs in the IBM replays compared to those in the UNC replay. There are of course other factors, like connection RTT, that also affect response times.

Although we discussed the *meanRTT* experiments in detail above, these results hold when using the *10pathRTT* or the *usernet* RTT models as well. That is, the *a-b* model has the fastest response times regardless of RTT model, because there are no intra-epoch latencies (endpoint latencies) within each epoch. The cross-over of the distribution of response times between the *block-sequential* and the *a-t-b-t* models in the IBM replays is probably due to the differences in impact of request and response sizes (larger in *block-sequential* model) vs. the impact of endpoint laten- cies (present in the *a-t-b-t* model) on epoch response times.

Figures 5.151 through 5.156 show the CCDFs for response times for these exper- iments. Clearly connection structure has a large effect on the distribution of response times even in the tails of the distributions, espcially considering that these figures are on a log-log scale. In all cases, regardless of the RTT emulation used, the *a-b* model has the lightest tail because it has the smallest ADU sizes, though same as that of the *a-t-b-t* but smaller than that of the *block-sequential*. The *a-t-b-t* model has the heaviest tail of response times because these epochs are dominated by the intra-epoch endpoint latencies for this connection structure model.

In the body of these distributions, we noted that the *a-b* model resulted in the shortest response times regardless of RTT model used. Also, the distribution of response times for the *a-b* model when in the UNC replay was heavier than in the IBM replay due to the smaller request and response sizes for IBM traffic. Those results still hold for the tails of these distributions across RTT models. However, the *a-t-b-t* model produces slightly heavier distribution of response times in the IBM replay than in the UNC replay, in the very end of the tail of the distributions. This is because the top 1% of intra-epoch latencies for connections in the IBM traffic was slightly heavier than that for connections in the UNC traffic.

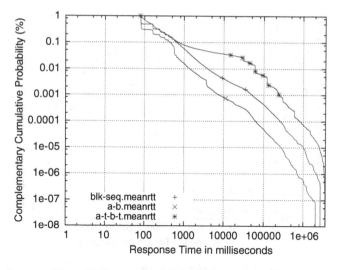

Fig. 5.151 Response Time – UNC (*meanrtt* round trip time)

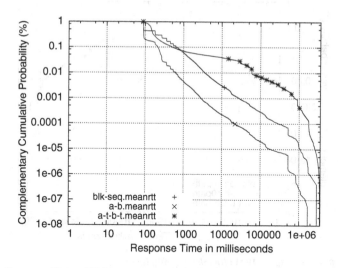

Fig. 5.152 Response Time – IBM (*meanrtt* round trip time)

5.3.3   *Effect of Structure on Queue Length at the Router*

In this section, we show the queue lengths at the outbound queue of the router before the *unconstrained* router-to-router link. The queue was sampled every 10 *ms* for the entire hour of the experiment. However, we only show the queue length data

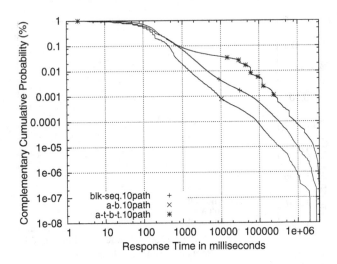

Fig. 5.153 Response Time – UNC (*10path* round trip time)

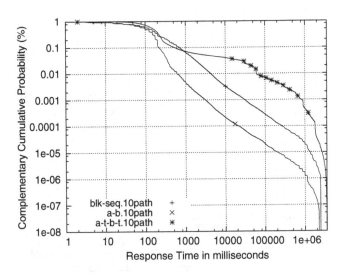

Fig. 5.154 Response Time – IBM (*10path* round trip time)

for the stable middle 40 minutes of the experiment. In Figs. 5.157 and 5.158, the distribution of queue length at the router's outbound queue is shown for four experiments in each of the UNC and IBM replays respectively. Each set of experiments used the *meanRTT* model while we varied the connection structure model per experiment among the *block-concurrent*, *block-sequential*, *a-b*, and *a-t-b-t* models.

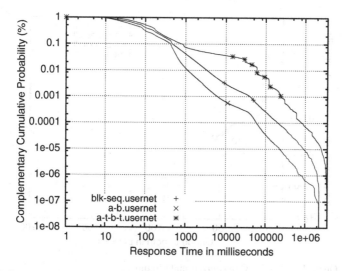

Fig. 5.155   Response Time – UNC (*usernet* round trip time)

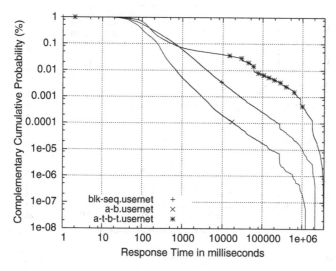

Fig. 5.156   Response Time – IBM (*usernet* round trip time)

Similarly, in Figs. 5.159 through 5.162, we show results for queue length for experiments varying the connection structure models while keeping the RTT emulation method the same between the *10pathRTT* and the *usernet* methods of emulation.

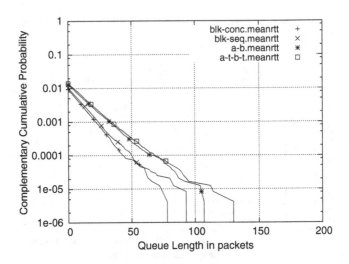

Fig. 5.157 Queue Length – UNC (*meanrtt* round trip time)

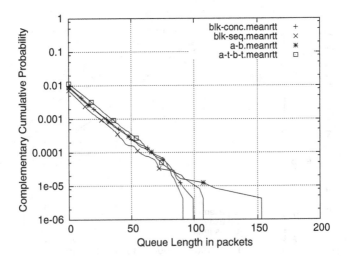

Fig. 5.158 Queue Length – IBM (*meanrtt* round trip time)

Figures 5.157 through 5.162 show the distributions of the queue length at the router. The average throughput in these experiments was around 471 Mbps for the UNC replays and 404 Mbps for the IBM replays. The router-to-router link was set to 1Gbps. Hence, for every combination of connection structure model and RTT emulation, the queue was empty for 99% of the time, as seen in these figures.

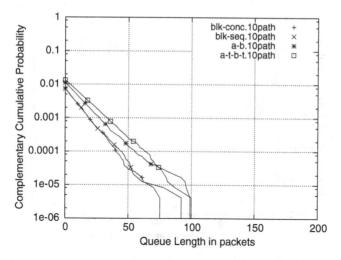

Fig. 5.159 Queue Length – UNC (*10path* round trip time)

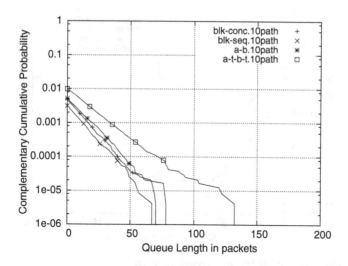

Fig. 5.160 Queue Length – IBM (*10path* round trip time)

The CCDFs show that the queue had up to 100 packets at times. This is because there were a few short periods of peak traffic arriving at the router queue at greater than 1Gbps from the 10Gbps aggregation link before the router. Hence, there were 10 or more packets in the queue for all the replays in the *unconstrained* mode for about 0.05% of the time.

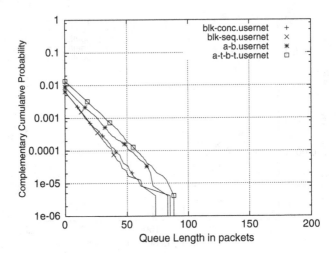

Fig. 5.161 Queue Length – UNC (*usernet* round trip time)

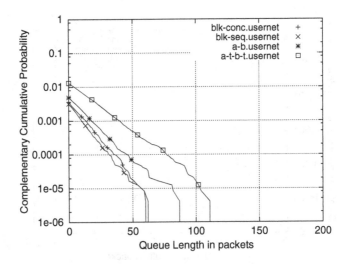

Fig. 5.162 Queue Length – IBM (*usernet* round trip time)

5.3.4   Effect of Structure on Active Connections

We define any TCP connection as an 'active connection' in the network at a given time t, if the SYN for that TCP connection has been seen on the network, but the FIN or RST has not yet been recorded. In this section, we study the effects on the number of active connections in the network when varying connection structure and keeping the RTT emulation method the same for that set. Figures 5.163 through 5.168 show

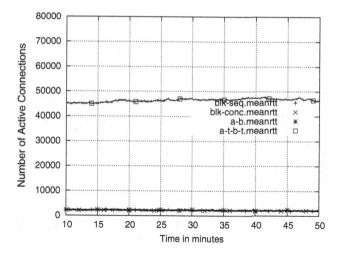

Fig. 5.163 Active Connections – UNC (*meanrtt* round trip time)

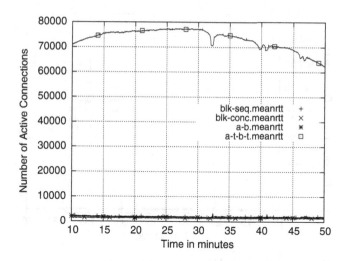

Fig. 5.164 Active Connections – IBM (*meanrtt* round trip time)

the time series of the number of connections that were recorded as active in the network during the middle 40 minutes of each experiment.

The *block-concurrent* and *block-sequential* models open the TCP connection and transfer bytes as quickly as possible. The *block-concurrent* model transmits data concurrently from both ends of the connection, while the *block-sequential* model transmits data sequentially, like one giant epoch per connection. The *a-b* model transmits data in epochs. However, all three models spend most of the connection

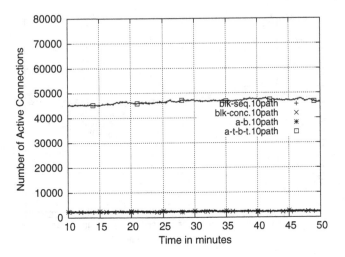

Fig. 5.165 Active Connections – UNC (*10path* round trip time)

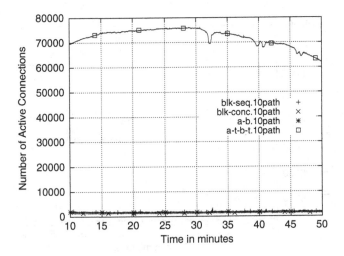

Fig. 5.166 Active Connections – IBM (*10path* round trip time)

duration in data transmission and RTTs, and hence they complete the connections very quickly.

The effect is that the number of active connections at any given time during the experiment is at least an order of magnitude lower for these three models as compared to the *a-t-b-t* model for both UNC and IBM replays, as seen in these figures. The *a-t-b-t* model preserves the endpoint latencies in each TCP connection. The number of active connections for the *a-t-b-t* model thus increases dramatically compared to the other three models.

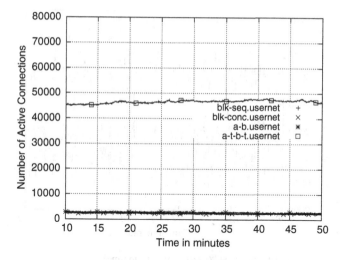

Fig. 5.167   Active Connections – UNC (*usernet* round trip time)

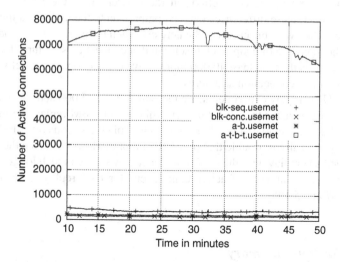

Fig. 5.168   Active Connections – IBM (*usernet* round trip time)

In Fig. 5.169, we change the y-axis but show the same data for the UNC replay as shown in Fig. 5.167. We observe here that there is indeed a difference in number of connections among the first three models, with *block-concurrent* having the least number of connections active in the network at any time, followed by *block-sequential* and then *a-b*. For example, at almost any given time during the experiment, the *block-sequential* model results in about an average of 200 more active connections in the network than the *block-concurrent* model. And the *a-b* model results in

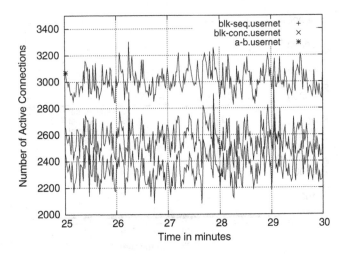

Fig. 5.169 Active Connections – UNC (*usernet* round trip time)

roughly 400 more active connections in the network than the *block-sequential* model. However, these differences pale in comparison to the multi-fold increase in the number of active connections in the network when using the *a-t-b-t* model. This is due to the modeling of endpoint latencies in that connection structure. The *a-t-b-t* model clearly results in at least 40,000 more active connections in the network than all the other three connection structures.

For the UNC replay experiments, we measured about 45,000 active connections throughout the hour for the *a-t-b-t* model, while for the IBM replay, we measured about 78,000 active connections. The total number of connections over the hour for UNC replay was almost double that of the IBM replay. However, since the connections in the IBM replay with the *a-t-b-t* model clearly showed much longer connection durations, the number of active connections for IBM replay is much higher (Figs. 5.167 and 5.168).

5.3.5   Section Summary

In this section, we presented results for replays in the *unconstrained* mode using UNC traffic and IBM traffic. We discussed the results for three sets of experiments for each of the two input traffic mixes. For each set of experiments, we kept the RTT model the same for all experiments, while varying the connection structure model from among the *block-concurrent*, *block-sequential*, *a-b* and *a-t-b-t* models. Thus we studied the effect of these empirically-derived connection structure models on four key performance metrics: connection duration, response time, router queue length, and active connections.

We found that the connection structure model used in emulating network characteristics has a significant impact on the performance metrics – orders of magnitude more than the effect of RTT emulation methods. The connection structure model significantly affects connection durations and response times both in the body and the tail of the distributions of these performance metrics. We also found that the router queue length showed no differences among the experiments using different connection structure models. This was expected because these were replays in the *unconstrained* mode, and hence designed to not create any queue buildup. The number of active connections in the network is a second order effect that is affected by connection durations. This metric was also greatly affected by the differences in the connection structure models used in the experiments.

So, if we had to choose a connection structure model to be used for experiments, which model would we pick? The choice of connection structure model is actually easier than the choice of RTT emulation. All the connection structure models we used were empirically derived from the same sources. But clearly, the *a-t-b-t* model with its endpoint latencies makes a huge difference in all outcomes for an experiment. So the take away message, if there is to be just one, is that the time components of traffic generation are as important as the size components. That is, while it is important to emulate TCP connections by the size of the connections, it is equally important to emulate them by the time components. These consist of the connection RTTs, the sequential or concurrent nature of data exchanges within connections, and the endpoint latencies measured for these connections. Unlike with RTT models, the connection structure models affect all the performance metrics significantly and throughout the distributions.

5.4 Effect of Connection Structure in the Constrained Mode

In Section 5.3 we discussed the effect of connection structure models on the four performance metrics of connection durations, response times, router queue length and active connections. Those were replays in the *unconstrained* mode; that is, the router-to-router link was set to 1Gbps. In this section we present results showing the impact of connection structure models on the same four metrics for a set of replays in the *constrained* mode; that is, the router-to-router link is set to 105% of the offered load on that link. For the replays in the *constrained* mode using UNC traffic, the link was set to 496Mbps, and for the replays in the *constrained* mode using IBM traffic, the link was set to 424Mbps. For each set of experiments, we compare the performance metrics for different connection structure models, keeping the RTT emulation method the same for all experiments in that set.

5.4.1 Effect of Structure on Connection Durations

Before we compare the effects of connection structure models on connection durations, we begin by studying the effect of the constraint on the router-to-router link on connection durations in the *constrained* mode. For this we compare the connection

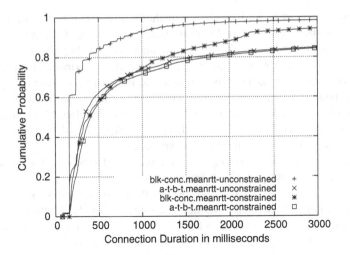

Fig. 5.170 Connection Duration – UNC (*meanrtt* round trip time)

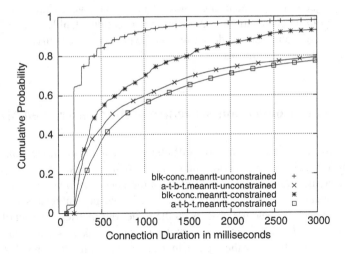

Fig. 5.171 Connection Duration – IBM (*meanrtt* round trip time)

durations for *block-concurrent* and *a-t-b-t* models in the *unconstrained* and *constrained* modes.

In Section 5.3.1, we observed the direct effect of connection structure modeling on the durations of connections when there was no constraint in the network. In this section, we observe that the connection duration is not only affected by the difference in connection structure, but it is even more significantly affected by the constraint on the link. In Figs. 5.170 through 5.175 we show the distributions of the connection durations for these experiments. In each figure, the RTT model is the

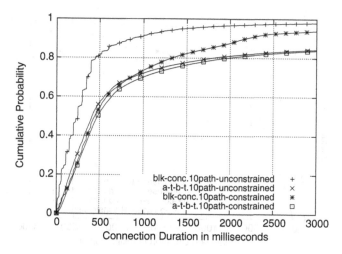

Fig. 5.172 Connection Duration – UNC (*10path* round trip time)

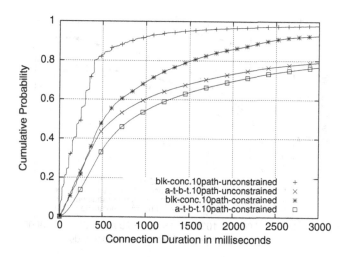

Fig. 5.173 Connection Duration – IBM (*10path* round trip time)

same for all experiments while the connection structure is varied. Each figure shows results from two replays in the *unconstrained* mode and two replays in the *constrained* mode. Each figure shows results from experiments using either the block-concurrent or the *a-t-b-t* connection structure models keeping the RTT emulation method constant using either the *meanrtt* or the *usernet* RTT emulation.

See Fig. 5.170 showing four UNC replay experiments. We find that 84% of connections complete in 500 *ms* or less using the *block-concurrent* model in the *unconstrained* replay, but only 58% of connections complete in the *constrained* replay for

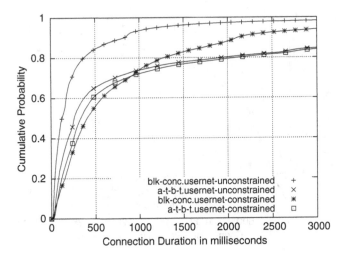

Fig. 5.174 Connection Duration – UNC (*usernet* round trip time)

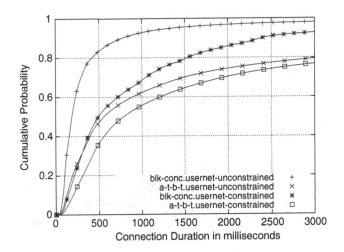

Fig. 5.175 Connection Duration – IBM (*usernet* round trip time)

the same connection structure. The effects of queuing delay are so profound on connection duration that for about 70% of connections, the duration is the same for the *a-t-b-t* model in the replay in *unconstrained* mode as for the *block-concurrent* model in the *constrained* case. That is, the queuing delay for the *block-concurrent* model in the *constrained* case is as large as the endpoint latencies that were present in the original connections (and represented in the *a-t-b-t* model).

This same effect is seen when using the *10pathRTT*, as shown in Fig. 5.172 for the UNC replay. In the case of *usernet* RTT emulation, Fig. 5.174 shows that the

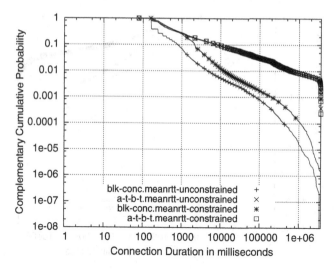

Fig. 5.176 Connection Duration – UNC (*meanrtt* round trip time)

queuing delays in the *block-concurrent* model exceed the endpoint latencies in the *a-t-b-t* model for 75% of connections, resulting in longer connections in the *block-concurrent* model than in the *a-t-b-t* model. As we will see in Section 5.4.3, the queue buildup for *block-concurrent* model is simply much heavier than for the *a-t-b-t* model. The other observation (Figs. 5.170, 5.172, and 5.174) is that for UNC replays, there is a very small difference in the distributions of connection durations between the *a-t-b-t* model in *unconstrained* and *constrained* modes. This is because the *a-t-b-t* model does not cause a huge queue buildup and hence the queuing delay is small, especially compared with the endpoint latencies that are the primary contributors to the connection durations.

Figures 5.171, 5.173, and 5.175 show results for IBM replay experiments in the *constrained* mode. The queue buildup in these experiments is slightly heavier for both *block-concurrent* and the *a-t-b-t* models as compared with the UNC replays. Hence we see large queuing delays affecting connection durations when using the *block-concurrent* connection structure model in the *constrained* mode. And even the *a-t-b-t* model creates a significant shift in the distributions of connection durations between its replays in the *unconstrained* and *constrained* modes. This is due to the heavier queue buildup in the IBM replay in *constrained* mode. Hence, we observe a greater difference in the distribution of connection duration between the *unconstrained* and *constrained* replays with the *a-t-b-t* model in the IBM replay than in the UNC replay.

Figures 5.176 through 5.181 show the CCDFs of connection durations for these experiments, varying connection structure while keeping the method of RTT emulation the same. Regardless of the RTT method used, there is still a large effect of the queuing delay on the connection durations for these long connections in the replays in the *constrained* mode using the *block-concurrent* model. With the *a-t-b-t* model,

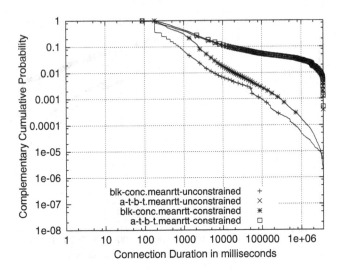

Fig. 5.177 Connection Duration – IBM (*meanrtt* round trip time)

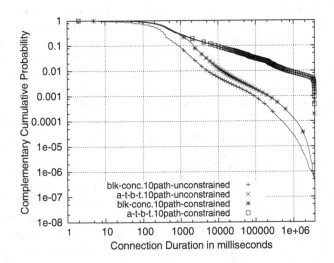

Fig. 5.178 Connection Duration – UNC (*10path* round trip time)

the queuing delay is masked by the much more dominant effect of the endpoint latencies within these very long connections.

So far, we have studied the difference in connection durations between replays in the *unconstrained* and *constrained* modes due to queuing delays caused by connection structure differences. We now present the results of replays in the *constrained* mode for studying the effect of connection structure modeling on

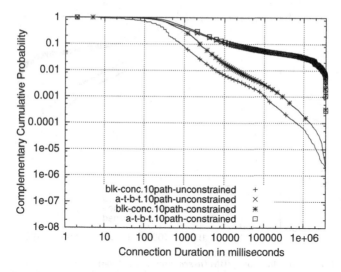

Fig. 5.179 Connection Duration – IBM (*10path* round trip time)

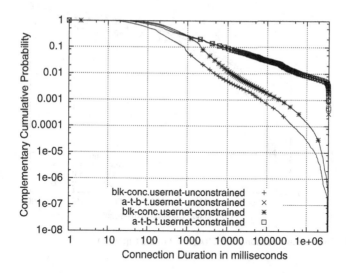

Fig. 5.180 Connection Duration – UNC (*usernet* round trip time)

connection durations. Figures 5.182, 5.184, and 5.186 show distributions of connection durations for the UNC replays in the *constrained* mode for four connection structure models. The connections in the *block-concurrent* and *block-sequential* models still complete faster overall than the other models but much slower than in their replays in the *unconstrained* mode.

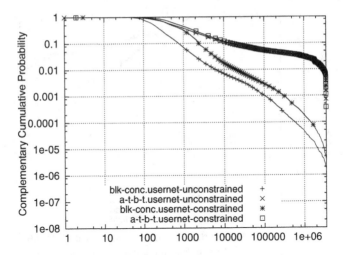

Fig. 5.181 Connection Duration – IBM (*usernet* round trip time)

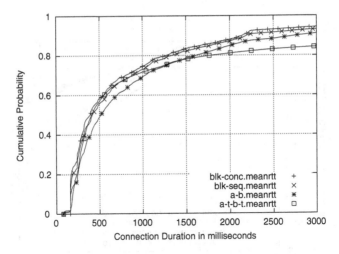

Fig. 5.182 Connection Duration – UNC (*meanrtt* round trip time)

The effects of queuing delay are so profound on connection duration that for the UNC replay experiments in the *meanRTT* and *10pathRTT* experiments, the connection duration for about 70% of connections is the same for the *a-t-b-t* model as for the *block-concurrent* and *blk-seq* models. That is, the queuing delay in the block models is as large as the endpoint latencies in the *a-t-b-t* model. In the case of *usernet* RTT emulation, Fig. 5.186 shows that the queuing delays in the block models exceed the endpoint latencies in the *a-t-b-t* model for 70% of the connections. Thus these connections take longer to complete in the block models than in the *a-t-b-t*

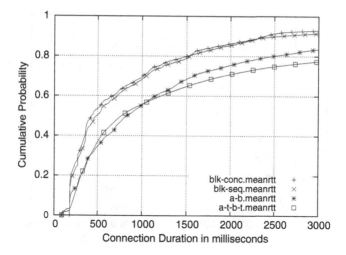

Fig. 5.183 Connection Duration – IBM (*meanrtt* round trip time)

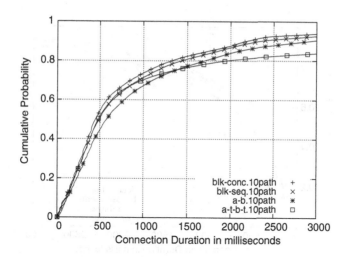

Fig. 5.184 Connection Duration – UNC (*10path* round trip time)

model. Recall that for a given connection structure, the *usernet* RTT model resulted in the longest queues.

Figures 5.183, 5.185 and 5.187 show results for the IBM replay experiments. The effect of queuing delay is seen in all four models of connection structure. We recall that even in the *unconstrained* mode, the *a-b* and *a-t-b-t* models in the IBM replays showed longer connection durations. This was due to the much larger number of epochs per connection in the IBM connections than in the UNC connections.

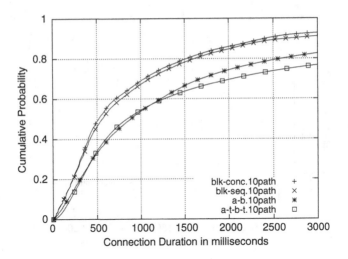

Fig. 5.185 Connection Duration – IBM (*10path* round trip time)

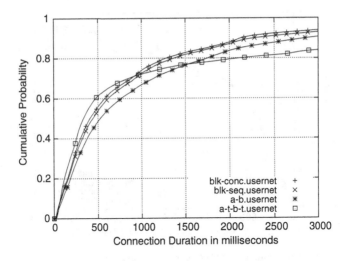

Fig. 5.186 Connection Duration – UNC (*usernet* round trip time)

So, now in the *constrained* mode, the replays with the *a-b* and *a-t-b-t* models continue to have the heavier distributions of connection durations. In the *constrained* mode, the effect of queuing delay further adds to the duration of these connections.

Figures 5.188 through 5.193 show the CCDFs of the connection durations for the different connection structures. We observe that for these long connections, the

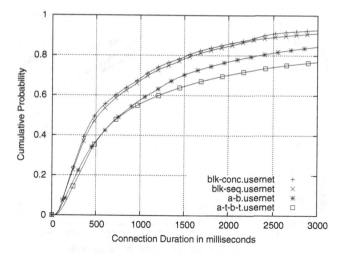

Fig. 5.187 Connection Duration – IBM (*usernet* round trip time)

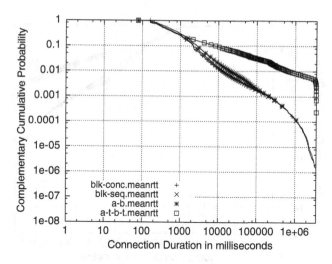

Fig. 5.188 Connection Duration – UNC (*meanrtt* round trip time)

connections using *block-concurrent* and *block-sequential* models complete at about the same rate. Connections using the *a-b* model take slightly longer due to the added time generating and transmitting data in epochs. The *a-t-b-t* model has the heaviest distribution of connection durations because of the significant effect of the endpoint latencies within these connections.

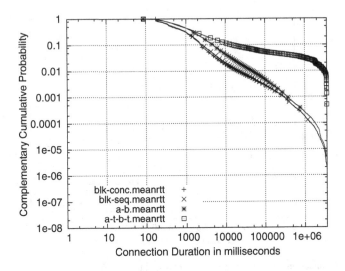

Fig. 5.189 Connection Duration – IBM (*meanrtt* round trip time)

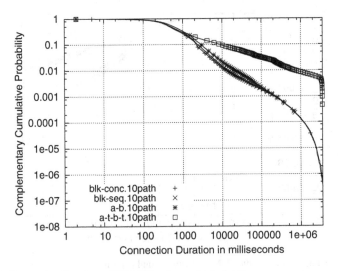

Fig. 5.190 Connection Duration – UNC (*10path* round trip time)

5.4.2   *Effect of Structure on Response Times*

In Section 5.3.2, we observed the direct effect of connection structure modeling on the response times for request-response exchanges in sequential connections when there was no constraint on the router-to-router link. In this section, we observe

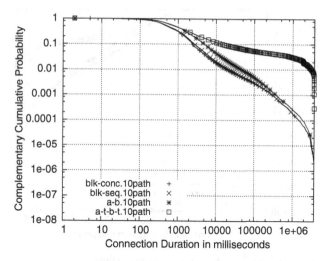

Fig. 5.191 Connection Duration – IBM (*10path* round trip time)

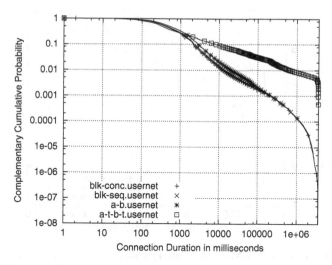

Fig. 5.192 Connection Duration – UNC (*usernet* round trip time)

response times not only affected by the difference in connection structure, but also
affected (even more significantly) by the second order effect of the queuing delay
that resulted from differences in the connection structure. Hence we first study the
effect of queuing delay for a combination of connection structure models and RTT
emulation methods. We compare the response times for the *a-b* and the *a-t-b-t*
models in the *unconstrained* and *constrained* cases in Figs. 5.194 through 5.199.

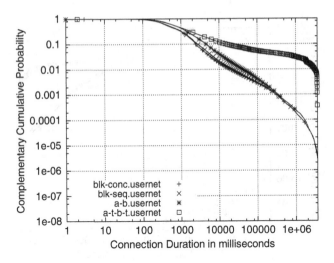

Fig. 5.193 Connection Duration – IBM (*usernet* round trip time)

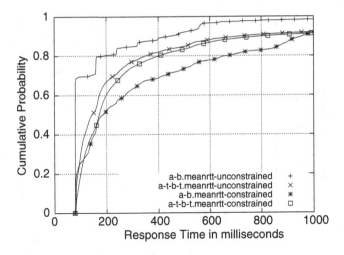

Fig. 5.194 Response Time – UNC (*meanrtt* round trip time)

The response time metric is most sensitive to differences in connection structure. Whereas the epochs in the *a-b* model experienced faster response times than the epochs in the *a-t-b-t* model in the replays in the *unconstrained* mode, the queuing delay in the replays in the *constrained* mode causes much longer response times for these same epochs. For example, in Fig. 5.194, while 90% of response times for the *a-b* model using *meanRTT* in the *unconstrained* experiment were 400 *ms* or less,

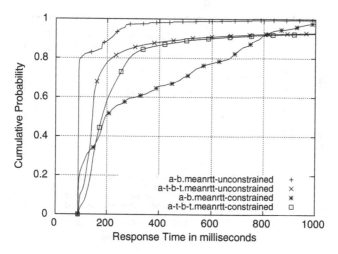

Fig. 5.195 Response Time – IBM (*meanrtt* round trip time)

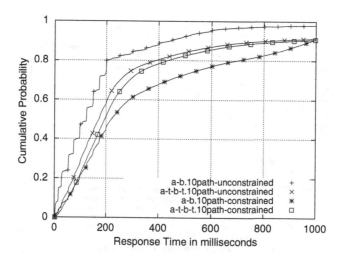

Fig. 5.196 Response Time – UNC (*10path* round trip time)

only 70% of these response times were less than 400 *ms* in the *constrained* experiment. For the *a-t-b-t* model with *meanRTT*, roughly 80-82% of response times were 400 *ms* or less for the replays in the *unconstrained* and *constrained* modes. The effect of queuing delay is slightly more pronounced for the *a-t-b-t* experiment in the *constrained* mode using IBM traffic as seen in Fig. 5.195.

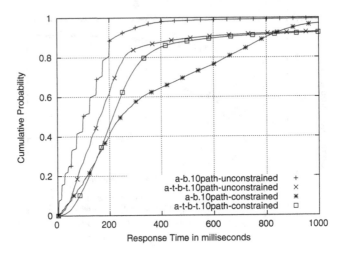

Fig. 5.197 Response Time – IBM (*10path* round trip time)

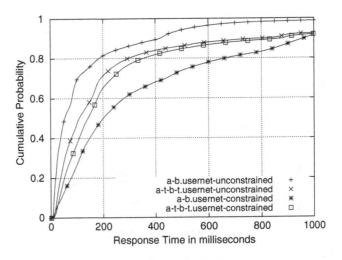

Fig. 5.198 Response Time – UNC (*usernet* round trip time)

Figures 5.196 and 5.197 show the response times for UNC and IBM replays using the *10pathRTT* model. Again, we see a very significant effect of queuing delay on the distribution when using the *a-b* model and relatively small effect of queuing delay when using the *a-t-b-t* model. Similar results are shown when using the *usernet* RTT model, as shown in Figs. 5.198 and 5.199.

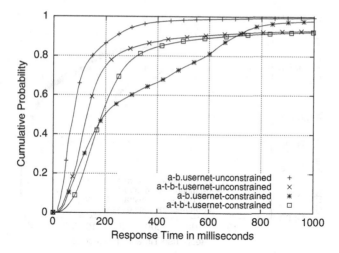

Fig. 5.199 Response Time – IBM (*usernet* round trip time)

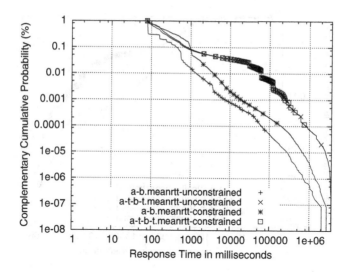

Fig. 5.200 Response Time – UNC (*meanrtt* round trip time)

Figures 5.200 through 5.205 show the CCDFs for the same experiments discussed above, showing the effect of queuing delay for the *a-b* and *a-t-b-t* models while keeping the RTT emulation method the same in each figure. We oberve that regardless of the RTT emulation method, there is a significant queuing delay effect on response times even for those long response times in the tails of the distributions.

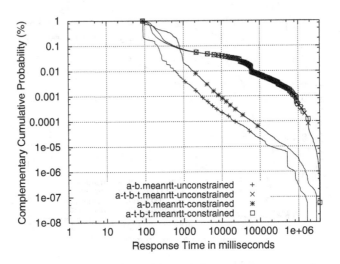

Fig. 5.201 Response Time – IBM (*meanrtt* round trip time)

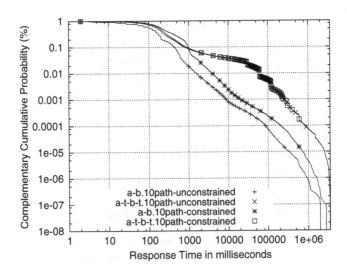

Fig. 5.202 Response Time – UNC (*10path* round trip time)

The *a-t-b-t* model shows no effect of queuing delay for response times in the long tail of the distribution. However, the response time distribution when using the *a-t-b-t* model is much heavier than those when using the *a-b-* model for replays in both *unconstrained* and *constrained* modes. This is because response times for the *a-t-b-t* model are dominated by the intra-epoch endpoint latencies which are orders of magnitude larger than the smaller queuing delays experienced by the connections in the *a-t-b-t* model.

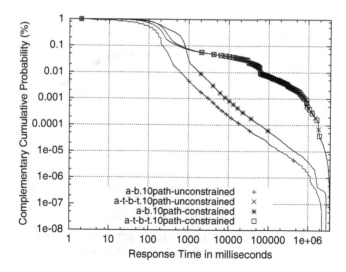

Fig. 5.203   Response Time – IBM (*10path* round trip time)

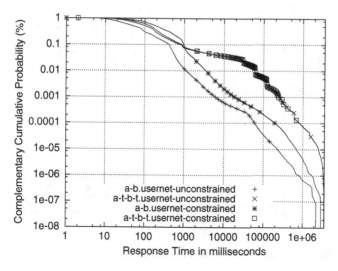

Fig. 5.204   Response Time – UNC (*usernet* round trip time)

So far, in this section, we have studied the effect of queuing delays on a given combination of connection structure and RTT models. We now compare the response times for different connection structure models for replays in the *constrained* mode. Figures 5.206 through 5.211 show response times for different connection structure models while keeping the RTT emulation method the same. We do not include the *block-concurrent* model in this section because that model has no notion

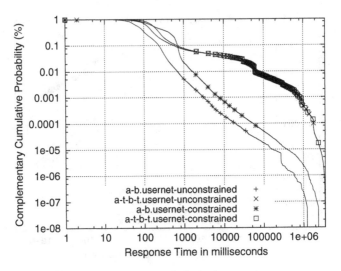

Fig. 5.205 Response Time – IBM (*usernet* round trip time)

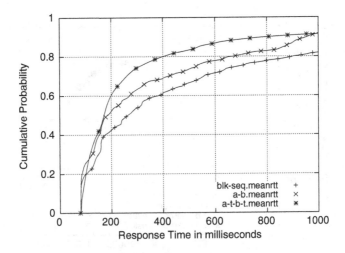

Fig. 5.206 Response Time – UNC (*meanrtt* round trip time)

of request-response exchanges and hence no notion of response times either. Overall, we observe from these figures that regardless of the RTT emulation method, the queuing delay has such a debilitating effect on response times for the *block-sequential* and *a-b* models that despite the *a-t-b-t* model generating intra-epoch endpoint latencies, the *a-t-b-t* connections show the fastest response times in the *constrained* cases for experiments using either UNC or IBM traffic. Interestingly, the same endpoint latencies within connections in the *a-t-b-t* model that were responsible

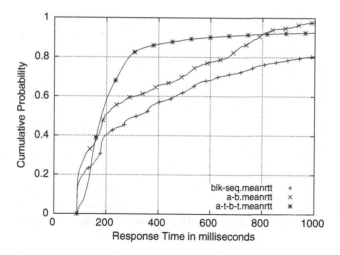

Fig. 5.207 Response Time – IBM (*meanrtt* round trip time)

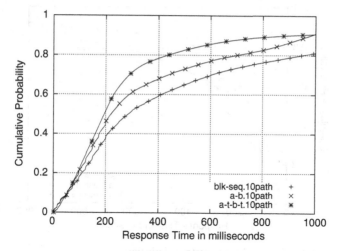

Fig. 5.208 Response Time – UNC (*10path* round trip time)

for longer response times in replays in the *unconstrained* mode are now also responsible for shorter queues and hence smaller queuing delays in the replays in the *constrained* mode; this leads to shorter response times for the *a-t-b-t* model compared to the *a-b* model in *constrained* mode.

In Fig. 5.206, we observe that 80% of epochs using the *a-t-b-t* model have response times less than 400 ms while only 68% of epochs using the *a-b* model have response times less than 400 ms. In the replays in the *unconstrained* mode, the response times of the *a-b* model were much shorter than those for the *a-t-b-t* model.

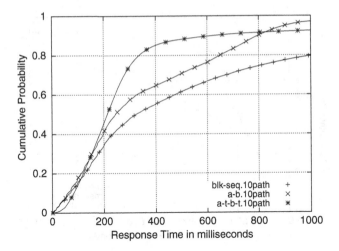

Fig. 5.209 Response Time – IBM (*10path* round trip time)

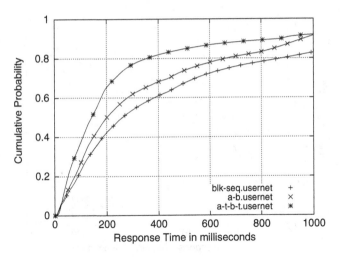

Fig. 5.210 Response Time – UNC (*usernet* round trip time)

This drastic shift in distribution of response time for the *a-b* model is due to very heavy queuing delays. Similarly, due to queuing delays, only 60% of epochs using the *block-sequential* model have response times less than 400 ms.

The difference in response time distributions among the different connection structures is significantly greater than the response time distributions among the different RTT emulation methods. Also, with both connection duration and response time, the effect of RTT emulation was seen up to about 500 *ms* to 1 second. The queuing delay was due to differences in connection structure more so than differences in RTT emulation. Hence we conclude that although RTT emulation affects

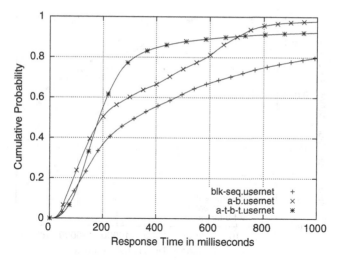

Fig. 5.211   Response Time – IBM (*usernet* round trip time)

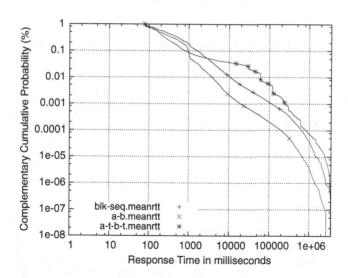

Fig. 5.212   Response Time – UNC (*meanrtt* round trip time)

end-user performance measures of connection durations and response times, the differences in connection structures, especially the endpoint latencies within TCP connections have the single most dominant effect on connection durations and response times, both due to the structure of the connections itself and due to the queuing delay effect of such structure.

Figures 5.212 through 5.217 show the CCDFs of the response times for the different connection structures while keeping the RTT emulation method the same in each figure. We observe that for these long response times, the request-response

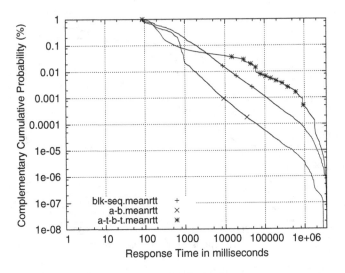

Fig. 5.213 Response Time – IBM (*meanrtt* round trip time)

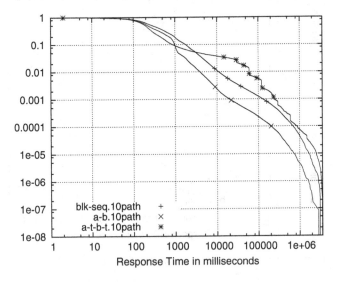

Fig. 5.214 Response Time – UNC (*10path* round trip time)

exchanges in the *a-b* model are the fastest since the ADU sizes of these single epochs is smallest along with having no endpoint latencies. The *block-sequential* connections have longer response times due to larger ADU sizes than the *a-b* model. The *a-t-b-t* model has the heaviest distribution of response times because of the significant effect of the intra-epoch endpoint latencies within these request-response exchanges.

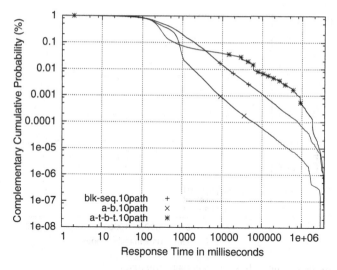

Fig. 5.215 Response Time – IBM (*10path* round trip time)

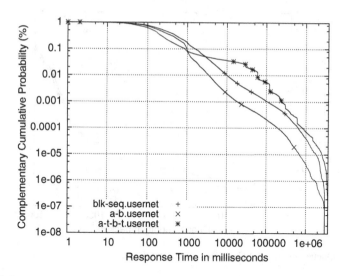

Fig. 5.216 Response Time – UNC (*usernet* round trip time)

5.4.3 Effect of Structure on Queue Length at the Router

In this section, we discuss the effects of different connection structures on the network-level performance measure of queue lengths at the outbound queue of the router before the *constrained* link. Figures 5.218 through 5.222 show the distribu-

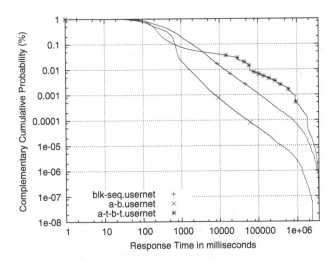

Fig. 5.217 Response Time – IBM (*usernet* round trip time)

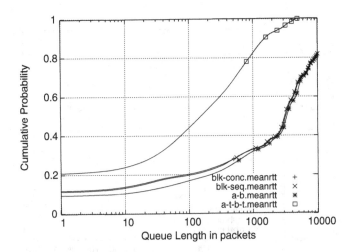

Fig. 5.218 Queue Length – UNC (*meanrtt* round trip time)

tions of the outbound queue length at the router, measured every $10\,ms$, during the middle 40 minutes in each hour long experiment. These figures represent the results from replays in the *constrained* mode. For the UNC replays the bottleneck link was set to 496Mbps and for the IBM replays it was set to 424 Mbps. The router queue was deliberately set to accommodate 65,000 packets so as not to cause any packet drops. The goal here was to determine the first order effect of different connection structure models on the router queue, and thus study the second order effects this had on connection durations, active connections and response times.

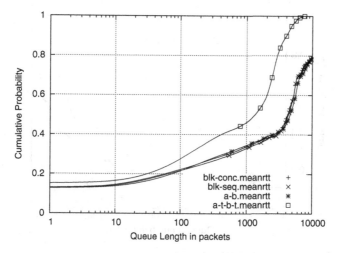

Fig. 5.219 Queue Length – IBM (*meanrtt* round trip time)

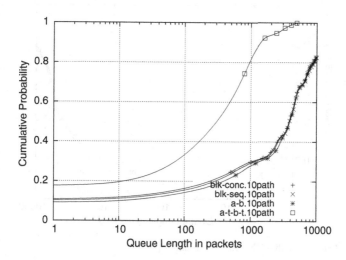

Fig. 5.220 Queue Length – UNC (*10path* round trip time)

With effectively 95% load on the link, and the moments of peak load creating even more queuing in the router, we see the router queues significantly loaded for much of the time. The IBM replay experiments show even greater queue occupation than the UNC replay experiments. This is directly due to the much higher load for IBM replay in the first half of the experiment since the original traffic had this characteristic of having greater throughput in the first half than in the second half of the traffic capture.

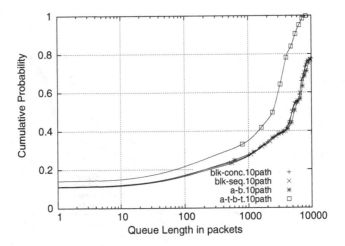

Fig. 5.221 Queue Length – IBM (*10path* round trip time)

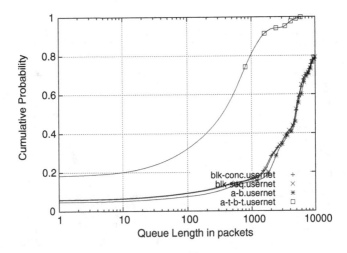

Fig. 5.222 Queue Length – UNC (*usernet* round trip time)

   In both sets of experiments using UNC and IBM traffic we observe that the
block-concurrent and *block-sequential* models do not allow the queue to drain for
most of the time. This is due to the back-to-back sending of windows of data packets
for connections using either of these models. In the block models, since there is no
separation of ADUs and no endpoint latencies between ADUs, the application can
send all the data at once, and TCP can grow its congestion window much faster. For
a given RTT, this leads to more packets outstanding in the network. Even the *a-b*
model, though it consists of epochs, does little or nothing to alleviate the queuing on
the router.

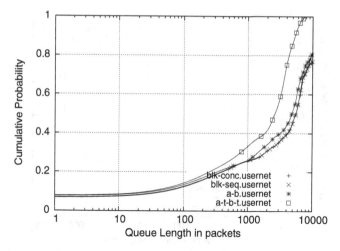

Fig. 5.223 Queue Length – IBM (*usernet* round trip time)

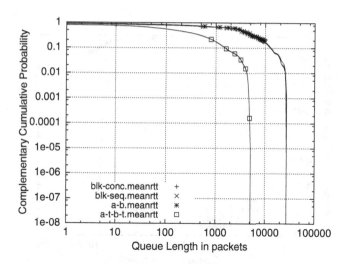

Fig. 5.224 Queue Length – UNC (*meanrtt* round trip time)

Thus for experiments using UNC traffic, these three models result in the queue having more than 1000 packets for 65% to 80% of the time, depending on RTT emulation method used. With the *a-t-b-t* model, however, the endpoint latencies in the connection structure allow the queue to drain and create different queue dynamics as a result. Only 20% of the time does the queue have more than 1000 packets in it. For the IBM replays, we see similar effects on the queue. However, the *a-t-b-t* model in this case does not alleviate the queue as much as in the UNC replays. This is because the endpoint latency distribution is much heavier in the UNC traffic than in the IBM traffic.

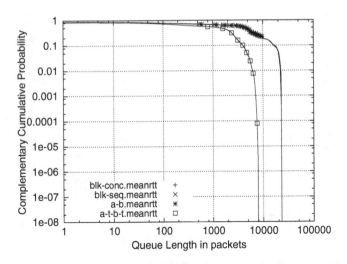

Fig. 5.225 Queue Length – IBM (*meanrtt* round trip time)

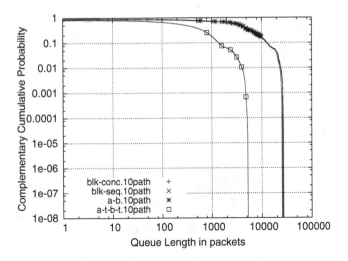

Fig. 5.226 Queue Length – UNC (*10path* round trip time)

5.4.4   Effect of Structure on Active Connections

In this section, we discuss the results for the number of active connections and compare the effect of connection structure models on this network-level measure in the *constrained* mode. Figures 5.230 through 5.235 show the time series of the number of connections that were seen active in the network for the middle 40 minutes of each experiment. The TCP connections using *block-concurrent, block-sequential,*

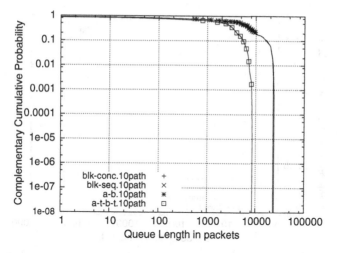

Fig. 5.227 Queue Length – IBM (*10path* round trip time)

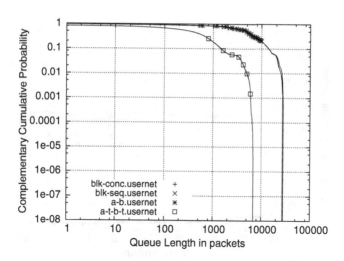

Fig. 5.228 Queue Length – UNC (*usernet* round trip time)

and *a-b* models spend most of the connection duration in data transmission, and hence they complete the connections very quickly. Hence, the number of active connections at any given time during the experiment is much lower for these three models as compared to the experiments using the *a-t-b-t* model. This is similar to results seen in the *unconstrained* cases.

The *a-t-b-t* model preserves the endpoint latencies in each connection. The number of active connections thus increases dramatically compared to the other three models. The active connections in the network are a direct consequence of connection durations experienced by the end user. One difference seen here, and not in the

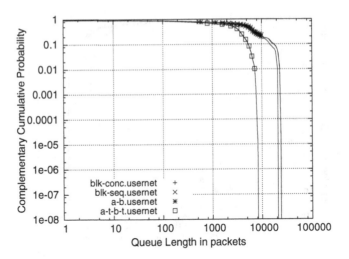

Fig. 5.229 Queue Length – IBM (*usernet* round trip time)

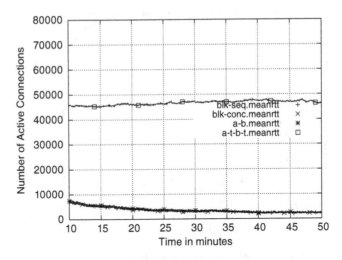

Fig. 5.230 Active Connections – UNC (*meanrtt* round trip time)

unconstrained case is the initial slightly higher number of active connections for the *block-concurrent, block-sequential* and *a-b* models. This was because of a much higher queuing delay experienced by these connections in the network during the initial several minutes of each experiment. Such queue dynamics are a direct consequence of unrealistically sending windows of packets back to back within a TCP connection ignoring all endpoint latencies that are an inherent part of the application models and hence connection structure. The lack of structure allows faster window growth, thus completing connections much faster.

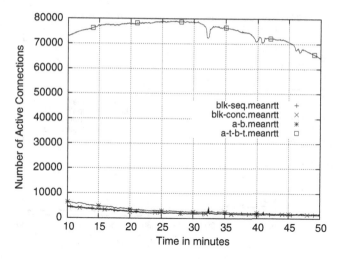

Fig. 5.231 Active Connections – IBM (*meanrtt* round trip time)

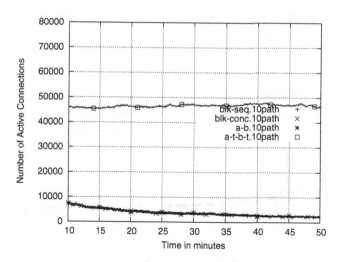

Fig. 5.232 Active Connections – UNC (*10path* round trip time)

5.4.5 Section Summary

For replays in the *constrained* mode, the connection structure model used for emulating TCP connections has a huge impact on connection durations. This is due to large queues and long queuing delays in the network which are a direct consequence of the connection structure used for traffic generation. There is also a significant impact on response times due to the connection structure model used for traffic generation.

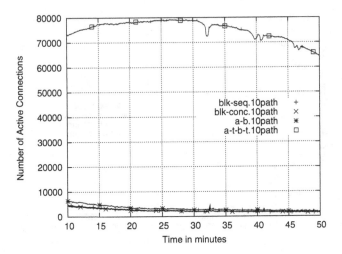

Fig. 5.233 Active Connections – IBM (*10path* round trip time)

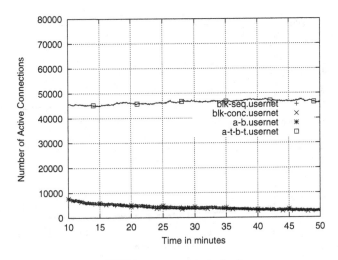

Fig. 5.234 Active Connections – UNC (*usernet* round trip time)

In this section, we also compared the effects of different connection structure models on the queue length at the router with the router-to-router link set to 105% of the offered load on that link, and a router queue length of 65,000 packets. Different connection structure models used in traffic generation affect queue dynamics differently. In general, the *block-concurrent* and *block-sequential* models create the heaviest queue length distributions because every connection sends all its bytes in one block. Thus in the absence of any latencies within the connection structure, the arrival pattern of packets at the queue is burstier for the block models than for the other

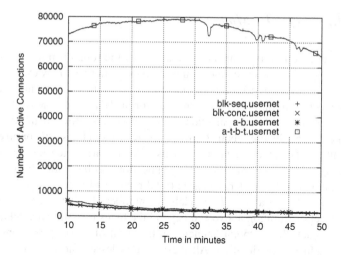

Fig. 5.235 Active Connections – IBM (*usernet* round trip time)

models. This is because for a given window size, the block and the a-b models are restricted only by the connection RTT in how quickly they can grow their window.

For connections using the *a-b* model, especially for those connections that originally had several epochs within the connection, the *a-b* model helps introduce latencies implicitly by way of generating synchronized request-response exchanges in sequence within the connection. Thus the replays in the *constrained* mode using the *a-b* model have a slightly lighter distribution of queue lengths compared with those using one of the block models. When connections are generated using the *a-t-b-t* model, they not only maintain the time sequence of request-response exchanges but also emulate intra-epoch and inter-epoch latencies within each TCP connection. This significantly alleviates queue buildup and hence the replays in the *constrained* mode using the *a-t-b-t* model result in the lightest queue distributions.

Hence we see the following patterns in all of the queue length distribution, regardless of the RTT emulation used. The heaviest queue length distribution is seen when using the *block-concurrent* and *block-sequential* models followed by the *a-b* model and finally the *a-t-b-t* model which creates the lightest distribution of queue lengths. All of these results hold true for the IBM replays in *constrained* mode as well. However, the queue length distributions in the IBM replay experiments were much heavier due to reasons discussed earlier, and directly related to the non-stationarity of the original throughput time series of the IBM traffic.

5.5 Chapter Summary

In this section, we summarize the effect of RTT models and connections structure models on network performance in 5.5.1 and 5.5.2 respectively.

5.5.1   Effect of RTT Models

In this chapter, we first presented our findings that the round trip time model used for traffic generation affects connection duration significantly. However, this effect is significant up to about 500 *ms* and, moderately so, up to 1 second in the distribution of connection durations. This holds true for all UNC and IBM replay experiments and for all the four connection structure models. Beyond the duration of 500 *ms*, and especially after about 1 second, the RTT model has little or no effect on the distribution of connection durations. This is due to various reasons. In the *block-concurrent* and *block-sequential* models, connections with duration beyond 500 *ms* are often dominated by the large filesize (total number of bytes transmitted) of the connection. For the *a-b* and *a-t-b-t* models, the number of epochs plays a significant role in connection duration above and beyond the bytes transmitted. And lastly, for the *a-t-b-t* model, the endpoint latencies in the form of server processing or user think times add to this duration. All these factors lessen the impact of the RTT model used in traffic generation, for durations beyond 500 *ms*.

So what do these sets of results tell us about how to emulate RTT for traffic generation for experiments where connection duration is a performance metric of importance? As long as the method of RTT emulation is representative of the empirical distribution of the traffic being replayed, the differences in RTT models affecting connection duration beyond 1 second are not significant. However, since the majority of connections in most production traffic are likely to have original connection durations less than 1 second, and since this is the region in which RTT emulation model matters most, if connection duration is an important performance metric in an experiment, then it would be advisable to use the *usernet* model for RTT emulation.

It is important to note that all three RTT methods discussed in this chapter are based on the same empirical measures and hence represent some form of realistic round trip times for TCP connections on the production link from which this traffic was acquired.

We also found that the round trip time of a connection significantly affects response times of epochs in that connection. This effect is seen mainly up to about 500 *ms* or 1 second of the response times. This is true for all experiments using both UNC and IBM traffic and holds for all the four connection structure models. Beyond the duration of 500 *ms*, or 1 second of the distribution, the RTT method has little or no effect on the distribution of response times. So which RTT model would we pick for traffic generation for experiments where response time is a performance metric of importance, as is often the case when evaluating new protocols or router queue mechanisms?

Since the majority of request-response exchanges in most traffic on production links are likely to have original response times much less than 1 second, and since this is the region in which RTT emulation model matters most, it would be advisable to use the *usernet* model of RTT emulation. Furthermore, the *meanRTT* or *10pathRTT* models lack the diversity of RTT values seen in the original distribution.

Thus the resulting distribution of response times for an experiment is varied when using the *usernet* RTT model but *constrained* to discrete values that are multiples of the few available connection RTTs when using *meanRTT* or *10pathRTT* models.

For experiments in the *constrained* mode, the RTT model had a significant impact on the distribution of queue lengths at the router. If there is a heavier distribution of connection RTTs in an experiment, that translates to more time on average between subsequent windows of packets in a TCP connection. The experiment using such an input RTT distribution in a *constrained* mode experiences more latencies within TCP connections, thus resulting in a lighter distribution of queue lengths because there is more time for the queue to drain. The number of active connections in the network is directly affected by the durations of connections in the network. We found that the RTT model does not affect the number of active connections in the network.

5.5.2 Effect of Connection Structure Models

The effect of RTT models on the application-level and network-level performance metrics, while significant, becomes almost negligible when compared to the dramatic effect of connection structure models on these metrics. While differences in RTT models clearly created differences in generation of the time component of each TCP connection, we found that the greater time components are actually part of the connection structure model.

The *a-b* model, even without any endpoint latencies can create a much heavier distribution of connection durations if there was a high average number of epochs in the original traffic being replayed. Modeling a TCP connection using the *a-t-b-t* model, which includes both epoch structure and endpoint latencies, captures all the original application data exchange patterns without knowledge of the actual applications. In doing so, this captures what we have discovered to be the most significant time component within TCP connections – the endpoint latencies.

We use connection duration as a performance metric not in the sense that the connection structure causing the fastest completions is the best model. Instead, our goal is to generate traffic such that the performance metrics measured during an experiment in the laboratory reflect the realistic measurements taken for connections in the original traffic. Why does this matter? Say, we develop a new transport protocol to operate at high speeds and enable faster connection completions. If we test this protocol in the laboratory using one of the block models, we cannot accurately assess whether faster connection completions are due to the block model or due to our new protocol. Hence, we conclude that if connection duration is a metric of importance, we should use the *a-t-b-t* model.

Response times for request-response exchanges within a TCP connection depend on three main factors: connection round trip time, the size of requests and responses, and the intra-epoch latencies. We see the clear effect of each of these components when studying the distribution of response times in this study. Connection RTT influences response times up to about 500 *ms* only. The sizes of requests and responses clearly have a significant effect on response times as seen from comparing the results between the *a-b* and the *block-sequential* connection structure models. And then intra-epoch latencies have the most significant effect on response times as seen from the results of the *a-t-b-t* model which generates all endpoint latencies.

The distributions of response times are often used as a metric of performance. For example, in Le et al. 2007, the authors demonstrate that one AQM scheme is better than another if the resulting distribution of response times when using the first AQM scheme in the routers is lighter (faster response times) than when using the other scheme. This is not the kind of assessment we seek to emphasize in this study. Faster response time for a connection structure in no way indicates that that connection structure is better than another. However, response time is an important metric in such protocol evaluation studies like the AQM example. And a metric is only good for comparison when it reflects reality; that is, when the traffic reflects the original request-response exchange sequence mimicking application behaviors found in the traffic on production links.

We found that the different connection structure models had significantly different effects on the queue length at the router. In general, the *block-concurrent* and *block-sequential* models create the heaviest queue length distributions because every connection sends all its bytes in one block. For connections using the *a-b* model, especially for those connections that originally had several epochs within the connection, the *a-b* model helps introduce latencies implicitly by way of generating request-response exchanges in sequence within the connection. Thus the replays in the *constrained* mode using the *a-b* model have a slightly lighter distribution of queue lengths compared with those using one of the block models. When connections are generated using the *a-t-b-t* model, they not only maintain the time sequence of request-response exchanges but also emulate intra-epoch and inter-epoch latencies within each TCP connection. This significantly alleviates queue buildup and hence the replays in the *constrained* mode using the *a-t-b-t* model result in the lightest queue distributions. The reason the tails of the response times seem unaffected is because the queuing delay, in the case of *a-t-b-t* connection structure experiments, represents a small fraction of the intra-epoch latencies measured for these connections in the original trace. Specifically, queuing delay is in tens of milliseconds while the intra-epoch latencies are hundreds of milliseconds to several seconds. For the top 10% of the epochs, response times in the *constrained* mode represent an increase of 10% and 20% for results for the UNC and IBM replays respectively as compared with their *unconstrained* modes.

Hence we see the following patterns in all of the queue length distribution, regardless of the RTT emulation used. The heaviest queue length distribution is seen when using the *block-concurrent* and *block-sequential* models followed by the *a-b* model and finally the *a-t-b-t* model creates the lightest distribution of queue lengths. All of these results hold true for the IBM replays in *constrained* mode as well.

The number of active connections in the network is directly affected by the durations of connections in the network. We observed that the connection structure model used in traffic generation significantly affects connection durations. The number of active connections in the network is a second order effect of the connection structure used in traffic generation. Thus we see that the number of active connections in the network is smallest when using the *block-concurrent* model and largest when using the *a-t-b-t* model, differing by an order of magnitude.

Reference

1. Feynman RP, Leighton RB, Sands M (1965) The Feynman Lectures on Physics. Addison Wesley, Boston

Chapter 6
Additional Results

> Discovery consists in seeing what everyone else has seen
> and thinking what no one else has thought.

Albert Szent-Gyorgi
(Hungarian Biochemist, 1937 Nobel Prize
for Medicine, 1893-1986)

So far in this book, we have presented results for a complete set of experiments using four connection structure models and three RTT models, in both *unconstrained* and *constrained* network link modes, using two different input traces. In this chapter, we present some additional results from experiments we conducted in the process of completing this study. For all experiments discussed in this chapter, we used only the UNC traffic as input. While these experiments are not central to our overall results, we have included them here for completeness. We present these results in three sections as follows. In Section 6.1, we present results for a small set of experiments using three RTT models we developed (explained in Chapter 3) – *nodelay*, *medianRTT*, and *uniformRTT*. For each of these RTT models, we ran experiments using only the control set (the *a-t-b-t* connection structure model with the *usernet* RTT model). In Section 6.2, we present results for the *DA* (*discrete approximation*) RTT model, showing that its results closely follow that of the *usernet* RTT model.

In Section 6.3, we present results from experiments varying another network-level parameter for traffic generation – receiver maximum window sizes assigned to endpoints of individual connections. Finally in Sections 6.4 and 6.5 respectively, we discuss the arrival patterns of packets at the router before the bottleneck link for different traffic models.

J. Aikat et al., *The Effects of Traffic Structure on Application and Network Performance*, 245
DOI 10.1007/978-1-4614-1848-1_6, © Springer Science+Business Media New York 2013

6.1   Miscellaneous Round-Trip Time Models

In Chapter 5, we presented results for experiments using three round trip time models for traffic generation for each of four connection structure models. They were *meanrtt*, *10pathrtt*, and *usernet* RTT. In this section, we show results for experiments using three other RTT models: the *nodelay*, *medianRTT*, and *uniform-RTT* models. For connection structure, we use only the *a-t-b-t* model, having already established that it most closely and realistically emulates the original traffic. For details on how we emulate these four models of RTT, we refer to Section 3.4: *Variations in emulating network path characteristics*.

6.1.1   Effect of RTT Emulation in the Unconstrained Mode

Let us begin this discussion with presenting results using these three RTT models in the *unconstrained* network mode. For each of the four performance metrics, we compare the results for experiments using *nodelay*, *medianRTT*, and *uniformRTT* models against experiments using the *usernet* RTT model as the control set. Sections 6.1.1 and 6.1.2 show results for experiments in the *unconstrained* and *constrained* modes respectively.

6.1.1.1   Connection Duration

In Figs. 6.1 and 6.2 we show CDFs and CCDFs for connection duration for four experiments – all using the *a-t-b-t* connection structure model, but different RTT

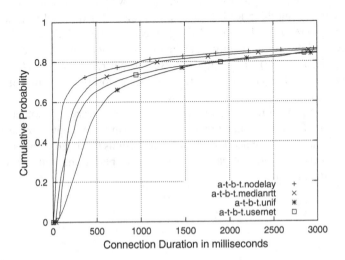

Fig. 6.1 Connection duration – CDF (a-t-b-t connection structure)

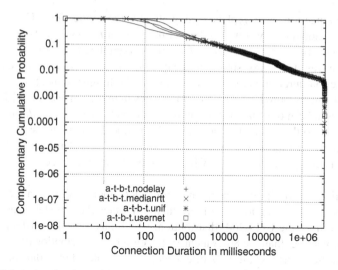

Fig. 6.2 Connection duration – CCDF (a-t-b-t connection structure)

models. As we observed with the three RTT models discussed in Chapter 5, we find that there is a significant difference in the distribution of connection duration among experiments using different RTT models. While the models used in Chapter 5 showed differences only up to 1 second, we observe a greater difference here. Why?

First, the *nodelay* model is really an extreme case where we emulate no connection RTTs at all. While this is not realistic, it serves a purpose here – to provide a quantitative assessment of the role of round trip times in connection duration. We find that although 60% of the connections complete in less than 127 *ms* with the *nodelay* RTT model, the mean completion time for connections using this model is still 33 *ms*. So, why do 40% of connections take more than 127*ms* to complete when there is no RTT delay? And why do 20% of the connections take more than 1 second to complete with no connection RTT? What is causing these connections to last so long? It is the epoch structure and endpoint latencies within the connections. We recall that 60% of the sequential connections had only one epoch, but almost 20% of sequential connections had 3 or more epochs for the UNC traffic which is the input traffic for all experiments discussed in this chapter. For these connections, almost 50% of these inter-epoch latencies were greater than 200 *ms* – much larger than most connection RTTs. In the case of concurrent connections, the endpoint latencies played an even more significant role in connection durations, with 60% of the endpoint latencies greater than 1 second. Hence although having no RTT delay makes a significant impact on the distribution of connection duration, it is not as dramatic as we would have expected, and this is because of the even more significant impact of endpoint latencies on the durations of connections.

The median RTT for all these 4.7 million connections in the input traffic was 36ms. Continuing discussion of the results shown in Fig. 6.1, we observe that when all connections were replayed using 36*ms* as the connection RTT – this is the *medianRTT* model – they completed faster than the *uniformRTT* and *usernet* cases. 60% of connections using the *medianRTT* model completed in 250*ms* or less while the average was still 33 ms, similar to the *nodelay* model. This is because the average connection duration is mostly influenced by the longer connections which are in turn influenced more by the endpoint latencies than the connection RTT. 20% of connections take more than 1.2 seconds to complete with the *medianrtt* model. Compare this to the results for the *usernet* RTT model which also shows an average connection duration of 33 ms. But with the *usernet* RTT model, 60% of those connections complete in 372 ms, while 20% of them take 2 seconds or more to complete.

Emulating round trip times using the *uniformRTT* model (shown in the results in Fig. 6.1) slowed completion time more than any other RTT model. This is likely because we used the U[10,200] distribution, which has a mean of 105ms. This is significantly higher than the 80*ms* mean of the empirical distribution of RTTs in the original traffic. Hence, while 60% of connections completed in 370*ms* or less with *usernet* RTT, it took up to 580*ms* for the completion for 60% of connections when using the *uniformRTT* model.

Comparing this result and those of the *10pathRTT* model (shown in Chapter 5) which is also a uniform distribution, we conclude that using a *uniformRTT* model is not necessarily an unrealistic method of RTT emulation. However, it is important to choose a distribution that has the same mean as that of the empirical distribution of connection RTTs for the particular set of connections being emulated. This is important for realistic traffic generation.

The CCDFs of connection durations, shown in Fig. 6.2, merely confirm what we have observed earlier with other RTT emulations; viz. that the effect of the RTT model used is greatly masked by endpoint latencies within connections, especially for connections with longer durations. Hence there is no difference in the tails of the distributions of connection durations using different RTT models.

6.1.1.2   Response Time

For the effect of RTT models on response times, we observe a similar trend as that of connection duration. That is, the fastest response times occur when no RTT delay is emulated (the *nodelay* model). In this case, while the *nodelay* model is obviously not a realistic RTT model, it again serves a purpose of differentiating between the impact of RTT delay and other latencies on epoch response times. 60% of response times are 68*ms* or less for the *nodelay* model as compared with 156*ms* or less when using the *usernet* RTT model. We find that response times are somewhat comparable for the *medianRTT* and *usernet*RTT models. The *uniformRTT* model results in the longest response times. Again, we attribute this to the longer average connection RTT for this model compared to the other three models for RTT emulation.

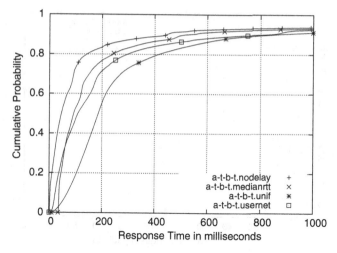

Fig. 6.3 Response Time – CDF (a-t-b-t connection structure)

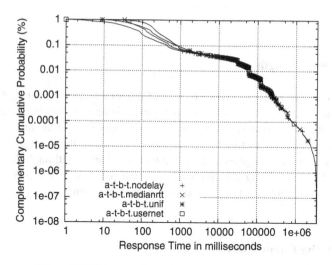

Fig. 6.4 Response Time – CCDF (a-t-b-t connection structure)

The difference in impact on response times due to different RTT emulations models diminishes after 1 second of response time distribution.

The distributions shown in Figs. 6.3 and 6.4 confirm that the RTT model used in an experiment has an effect on response times up to 1 second, but not beyond that. For response times greater than 1 second, the ADU sizes and intra-epoch latencies play a more significant role than the RTT model used for traffic generation.

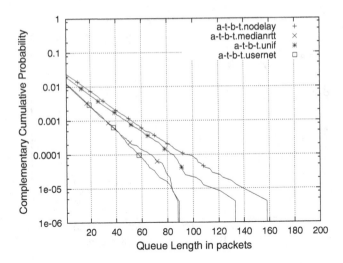

Fig. 6.5 Queue Length – CCDF (*a-t-b-t* connection structure with different RTT emulations)

6.1.1.3 Queue Length

The results for queue length, shown in Fig. 6.5 are for experiments run in the *uncon-strained* mode. Hence, as was observed for experiments shown in Chapter 5, we find that in the *unconstrained* mode, the queue is empty most of the time, except for the occasional spikes in traffic. In this *unconstrained* mode, the router-to-router link is 1Gbps, and mean throughput is less than half that. Hence, although the change in RTT model creates changes in the packet arrival patterns on the link before the router, the throughput is not high enough to cause queue buildup.

6.1.1.4 Active Connections

The number of active connections in the network is a direct result of the duration of connections in the experiment. So we find that the *nodelay* and *medianRTT* models result in the least number of active connections, relatively speaking. And the *uniformRTT* model results in the maximum number. However, as shown in Fig. 6.6, since all these experiments use the *a-t-b-t* model of connection structure, the number of active connections is about 45,000 connections, which is fairly high regardless of RTT model used, but this is due to the endpoint latencies within these connections.

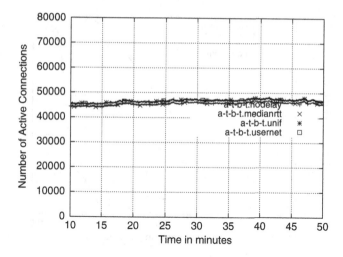

Fig. 6.6 Time series of Active connections (a-t-b-t connection structure)

6.1.2 Effect of RTT Emulation in the Constrained Mode

In this section, we discuss results for experiments using the UNC traffic in the constrained mode. In all experiments discussed here, we use the *a-t-b-t* connection structure model while varying RTT among the *nodelay*, *medianrtt*, *uniformrtt*, and *usernet* models.

6.1.2.1 Queue Length

We observed in Chapter 5 how a heavier RTT distribution caused a lighter router queue length distribution as a result of using that RTT model. That observation continues to hold true for the RTT models discussed in this chapter. Figures 6.7 and 6.8 show the queue length distributions for the four RTT models discussed in this section. The *nodelay* model is obviously the lightest RTT distribution, and hence creates the longest queue lengths. We observe that while the *usernet* RTT distribution causes the queue to have 1000 or more packets for about 20% of the time, the *nodelay* model causes the queue to have 1000 or more packets for 93% of the time. Thus queuing dynamics are drastically affected when the RTT delay is smaller. That is, as seen before, larger RTTs on average lead to more time between packets and thus chances for the queue to drain. *Nodelay* results in quick and massive queue buildup that takes a long time to drain. The effect of the queuing delay induced by this buildup was seen in both connection duration and response time.

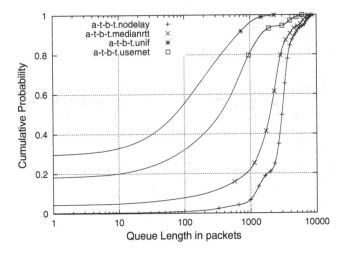

Fig. 6.7 Queue Length – CDF (a-t-b-t connection structure)

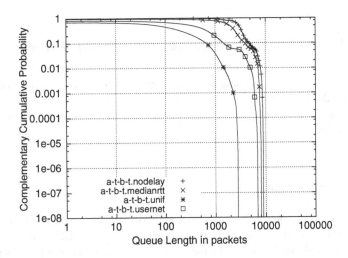

Fig. 6.8 Queue Length – CCDF (a-t-b-t connection structure)

MedianRTT emulation means that every connection has only 36*ms* round trip time, and hence connections that originally had much longer RTTs now send windows of packets much faster back-to-back into the network, causing queue buildup. The queue barely drains and has 1000 packets or more for fully 78% of the time. With *usernet*, there is a wide range of RTT delays, the average being 80ms, which is much longer than the median RTT of 36ms. Hence there are 1000 or more packets for only 20% of the time. More time to drain the queue means less impact of queuing delay on connection duration and response times, as observed in the previous sections.

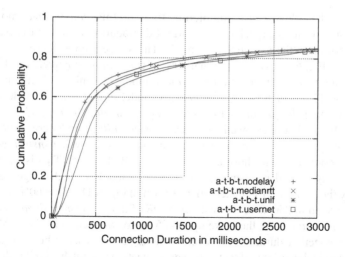

Fig. 6.9 Connection duration - CDF (a-t-b-t connection structure)

The *uniformRTT* model, with even larger average connection RTTs, results in 1000 or more packets for only 4% of the time. So, there is little queuing delay compared to the other RTT models. Still, the connection duration and response times were longer than for the other RTT models with larger queues. Why? That was due to the longer connection RTTs, and not due to added queuing delay with this emulation.

The CCDFs for the queue lengths in Fig. 6.8 show fairly significant differences in queue buildup for the different RTT models. Uniform RTT shows the lightest tail, followed by *usernet*, then *medianRTT* and finally the *nodelay* model. The top 10% of the distributions show that the queue has 650 packets or more when using the *uniformRTT* model, 1460 packets or more when using the *usernet* RTT model, 3869 packets or more when using the *medianRTT* model, and 4710 packets or more when emulating *nodelay* as the RTT model in the experiment.

6.1.2.2   Connection Duration

We now discuss results from experiments using the same set of connection structures and RTT models discussed in Section 6.1.1, but with the router-to-router link in *constrained* mode. In this mode, the link capacity is set to 105% of the mean throughput on that link. As we recall from the results in Chapter 5, the *a-t-b-t* connection structure model does not cause as severe a queue buildup as the other connection structure models, regardless of RTT emulation method used in the experiments. This was mostly due to the endpoint latencies within these connections that allowed the queue to drain between packet arrivals. Why is this significant? Because queue buildup causes queuing delay which increases connection duration. Hence we observe in Fig. 6.9 that while the *nodelay* RTT model still has the fastest completion time in this *constrained* mode, the connection durations experience the largest degradation from their corresponding distribution in the *unconstrained* mode. This is seen more clearly in Fig. 6.11 and 6.12. In Fig. 6.11, we observe that 60% of connections completed in 125*ms* or less with the *nodelay* model in *unconstrained* mode, but only 21% of connections complete in 125*ms* or less with the *nodelay* model in *constrained* mode. 60% of connections actually take up to 400*ms* to complete in *constrained* mode.

Similarly, while the connections using the *medianRTT* model completed much faster in the *unconstrained* mode, these completion times, now in *constrained* mode, are just as comparable to those using the *usernet* RTT model. This is because the queuing delay caused by using the *medianRTT* model is much longer than that caused by the *usernet* RTT model. So, as shown in Fig. 6.11, while 60% of connections completed in 250*ms* or less with *medianRTT* in the *unconstrained* mode, now with the *constrained* link they take up to 485 *ms* to complete. This is up 94%. The average connection duration, however, remains 33 ms. This is because the RTT emulation model does not affect the longer connections as much, and it is the longer connections that skew the mean duration.

Finally, the *uniformRTT* model again results in the longest completion times as seen in Fig. 6.12. However, the change in the distribution of connection durations between the *unconstrained* and *constrained* modes for this RTT model is small compared with the others. For instance, 60% of the connections completed in 588 *ms* in the *unconstrained* mode, and 621 *ms* in the *constrained* mode when using the *uniformRTT* model. This is because the average of this distribution of RTTs is higher, and as we mentioned earlier, the heavier the RTT distribution, the lighter is the queue length caused by the traffic, all else remaining the same. Hence while we saw a 94% increase in duration for the shortest 60% of connections between the *unconstrained* and *constrained* modes when using the *medianRTT* model, we only observed a 6% increase in duration using this *uniformRTT* model. This is not due to the uniformity of the distribution, but rather due to the larger mean for this *uniformRTT* model.

The CCDFs of connection duration shown in Fig. 6.10 confirm that either the endpoint latencies or the effect of queuing delays add up to overshadow any effects of RTT models for very long connections in all these cases.

6.1.2.3 Response Time

Figures 6.13 and 6.14 show the results in CDFs and CCDFs for response times for experiments using the four different RTT models. Again, we see the second order effect of queuing delay on response times. That is, in the *unconstrained* mode, different RTT models created different response time distributions purely due to differences in connection RTTs. But now in the *constrained* mode, there is the added

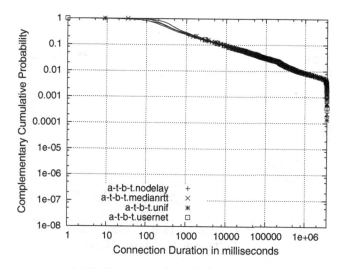

Fig. 6.10 Connection duration - CCDF (a-t-b-t connection structure)

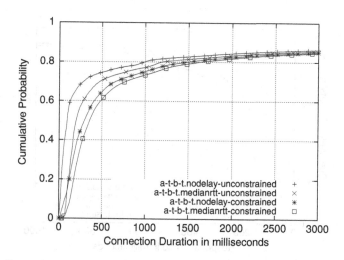

Fig. 6.11 Connection duration – CDF - UNC (a-t-b-t connection structure)

effect of queuing delay, and this queuing delay is different for the different RTT models. What do we mean by this?

Consider the *nodelay* RTT model. 60% of response times were 68*ms* or less for the *nodelay* model as compared with 156*ms* or less when using the *usernet* RTT model in the *unconstrained* mode. But in *constrained* mode, where the *nodelay* RTT model causes very large queuing delays, we observe that 60% of response times actually take up to 156 ms, which is a 129% increase in response times, whereas

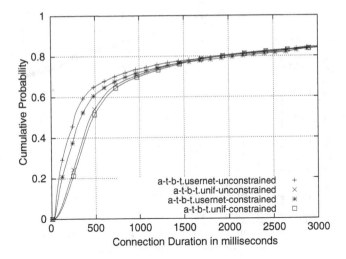

Fig. 6.12 Connection duration – CDF - IBM (a-t-b-t connection structure)

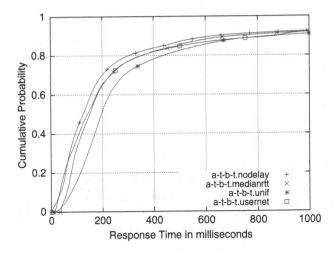

Fig. 6.13 Response Time – CDF (a-t-b-t connection structure)

60% of response times in the *usernet* model is now 178 *ms* which is a 14% increase in response time.

Similarly, the response time distribution when using *medianRTT* is much heavier than its corresponding distribution in the *unconstrained* mode. This has led to a decrease in what was previously a significant difference in the distributions for response times for the four different RTT models. Indeed, *medianRTT* seems to mirror the effect of *usernet* RTT, but that is only because the queuing delay from using

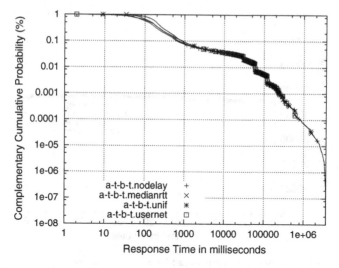

Fig. 6.14 Response Time – CCDF (a-t-b-t connection structure)

the *medianRTT* model has caused a significant increase in response times in the *constrained* mode.

Finally, the *uniformRTT* model has the longest response times, but this distribution is very similar to that obtained when using the *uniformRTT* model in the *unconstrained* mode. The fastest 60% of response times increased from 215 *ms* to 229 *ms* which is a mere 6% increase. Figure 6.14, using CCDFs for the response times, clearly shows that the RTT model used has no effect beyond 1 second in the distribution of response time.

6.1.2.4 Active Connections

The small differences in the number of active connections that we observed in the *unconstrained* mode, when using these different RTT models, are now overshadowed by the second order effect of queuing delay in the *constrained* mode. What do we mean? Number of active connections in the network is directly dependent on connection durations. The differences in connection durations due to the different RTT models reduced due to the longer queuing delays in the *nodelay* and *medianRTT* models. Hence, we observe, as shown in Fig. 6.15, that there is not much difference in the number of active connections in the network among the four RTT models.

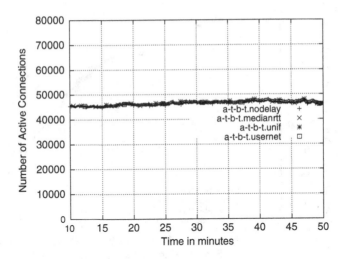

Fig. 6.15 Time series of Active connections (a-t-b-t connection structure)

6.2   Discrete Approximation (DA) RTT

Now we examine the effect of another RTT model – the Discrete Approximation (DA) RTT model, also called the 30path model. We developed this model as an approximation of the cumulative distribution of RTTs seen in the original trace. Using the concept of a quantile function (see Chapter 3 for more details), we approximated the CDF of the empirical RTTs as follows: we divided the distribution into 30 bins, and then found the average RTT for each of these 30 bins in the distribution. The resulting RTT values formed this set: [8, 8, 10, 10, 12, 14, 14, 16, 18, 20, 22, 24, 26, 30, 34, 38, 42, 48, 52, 60, 74, 80, 82, 86, 92, 98, 124, 172, 258, 420] milliseconds.

The reason we discuss this model separately is that the results from this RTT emulation most closely resemble the results using the *usernet* RTT model. Hence, we present this model as a realistic and reliable approximation for the standard *usernet* model. Why does this matter? Emulating *usernet* involves measuring every connection RTT and assigning the original connection RTT to that exact connection at the time of traffic generation. The *DA* RTT model is an approximation of the empirical model and is easier to implement because it only requires that we pick a discrete set of values that approximate the original minimum RTT distribution, and then assign these values to a small number of end-to-end paths in the experimental network. Hence, where appropriate, the *DA* model could be used for realistic RTT emulation.

6.2.1   Results in Unconstrained Mode

In this section, we present the results for all four performance metrics using the *DA* model in the *unconstrained* mode. For comparison, we show results for the control or *usernet* model of RTT. Both experiments were run using the *a-t-b-t* connection

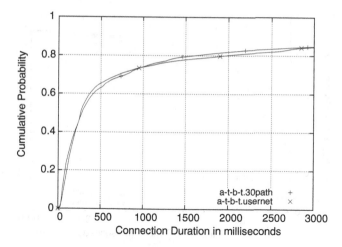

Fig. 6.16 Connection duration - CDF (a-t-b-t connection structure)

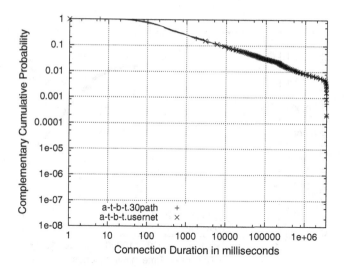

Fig. 6.17 Connection duration – CCDF (a-t-b-t connection structure)

structure model. In Figs. 6.16 through 6.21, we show the CDFs and CCDFs for connection duration, response time, and queue length for these two RTT models. In Fig. 6.22, we show the time series of the number of active connections during these experiments.

The distribution of connection duration for the *DA* RTT model practically tracks that of the *usernet* RTT model for the body as well as the tail of the distributions – see Figs. 6.16 and 6.17. Figures 6.18 and 6.19 show that the distribution of response time for the *DA* RTT model also closely tracks that of the *usernet* model for the

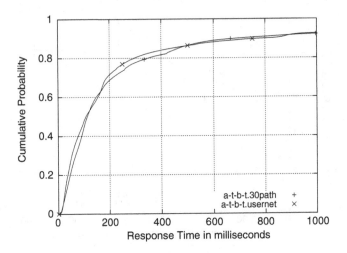

Fig. 6.18 Response Time – CDF (a-t-b-t connection structure)

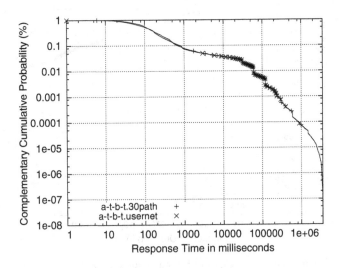

Fig. 6.19 Response Time – CCDF (a-t-b-t connection structure)

body and the tails of these distributions. For both connection duration and response times, it is to be expected that the tail of the distributions would remain the same, since we already observed that the RTT model used in traffic generation does not affect these metrics beyond 1 second in most cases, and up to a maximum of 3 seconds in some of the models discussed in this chapter.

However, for shorter connection durations and response times, there were significant differences among the different RTT models studied so far. Hence, it is

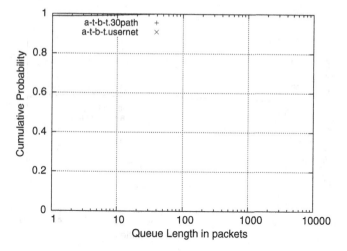

Fig. 6.20 Queue Length – CDF (a-t-b-t connection structure)

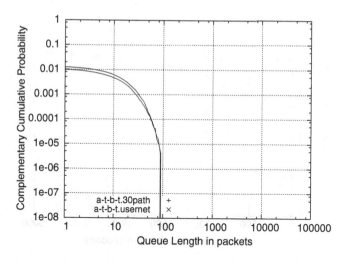

Fig. 6.21 Queue Length – CCDF (a-t-b-t connection structure)

noteworthy that of all the other six RTT models we developed and tested, none of them matched the control *usernet* model as closely as this *DA* model. The results for the queue length distributions as well as the time series of active connections in the network are also very similar when using the two RTT models in the *unconstrained* mode. So, clearly, if these were the performance metrics of interest, then the *DA* model could work just as well for RTT emulation as the *usernet* model.

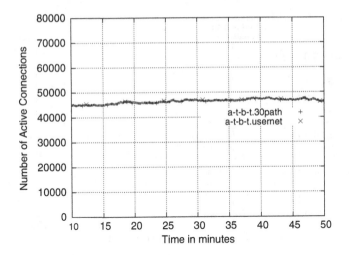

Fig. 6.22 Time series of Active connections (a-t-b-t connection structure)

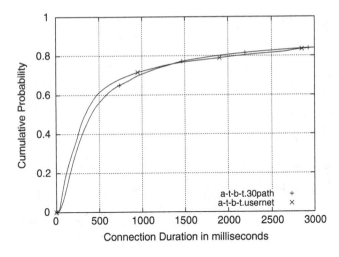

Fig. 6.23 Connection duration - CDF

6.2.2   Results in Constrained Mode

In this section, we present the results for all four performance metrics using the *DA* model in the *constrained* mode. Figures 6.23 through 6.29 show the CDFs and CCDFs for connection duration, response time, and queue length for the *DA* and the

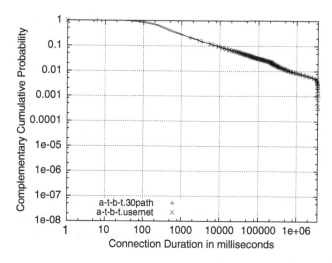

Fig. 6.24   Connection duration - CCDF

usernet RTT models. Finally, in Fig. 6.29, we show the time series of the number of active connections during these experiments.

Even in the *constrained* mode, the *DA* RTT model results in a distribution for connection durations that is comparable to that when using the *usernet* model. For response times below 500ms, as shown in Fig. 6.25, there is a small shift, with *usernet* having faster response times. This is due to the fairly large difference in queue buildup for the *DA* model as compared with the *usernet* model, as seen in Fig. 6.27. This buildup for the *DA* RTT model is likely due to the fact that RTT emulation by paths (which is what the *DA* model emulates) would lead to many connections that originally had long RTTs now having very small RTTs (and vice versa). In such cases, if these connections also had a large amount of data to send, then that would directly and drastically affect the queue.

Even so, we find that the effect of this fairly significant difference in queuing dynamics is not as large on connection duration and response time as might be expected. Note that the 67 *ms* mean RTT of the *DA* RTT model is less than the 80 *ms* mean RTT of the *usernet* model. As seen before, the smaller the mean of the RTT distribution, the longer is the queue at the router. The number of active connections is the same for both RTT models throughout the experiment as shown in Fig. 6.29.

The CCDFs shown in Figs. 6.24 and 6.26 show that the connection duration and response times produce similar tails in their distributions for these two RTT models. In Fig. 6.27 showing the tail of the queue length distributions, we observe that the two RTT models cause similarly long tails for the top 10% of the time.

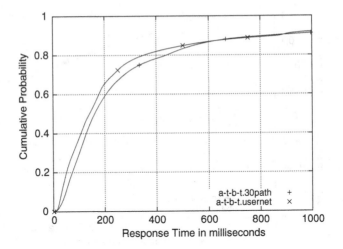

Fig. 6.25 Response Time – CDF

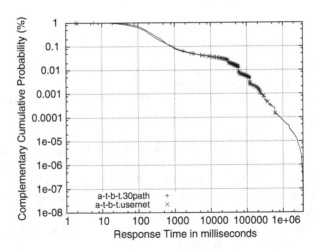

Fig. 6.26 Response Time – CCDF

6.2.3 Section Summary

The discrete approximation model of RTT emulation may be used instead of the *usernet* RTT model where a simpler yet empirically-based model is desired for traffic generation. If the experiment does not involve heavily congested scenarios, the *DA* RTT model produces results at the application-level and network-level that are very similar to that of the *usernet* model. In the presence of heavily *constrained* links in the network, the *DA* model creates significantly longer queues, and hence this must be taken into account when using this model for traffic generation.

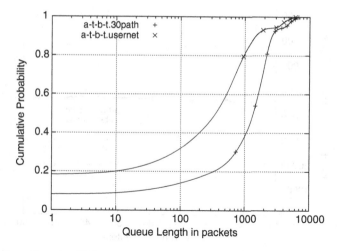

Fig. 6.27 Queue Length – CDF

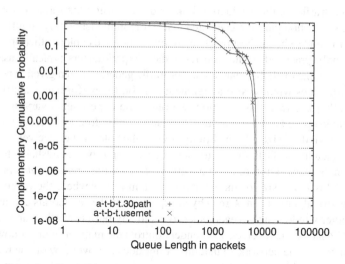

Fig. 6.28 Queue Length – CCDF

6.3   Emulating Receiver Window Sizes

In all the sections in Chapter 5 and the previous sections in this chapter, we discussed results for using all combinations of the four connection structure models and seven RTT models we developed for realistic traffic generation. In every experiment so far, we used the Tmix model of window size assignments to connections. That is, we measure the window sizes for each of the two endpoints for every connection in

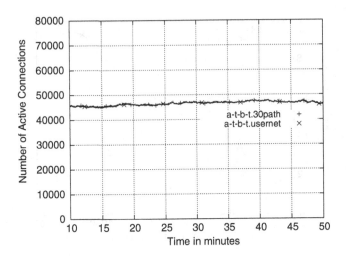

Fig. 6.29 Time series of Active connections

the original traffic on the production link, and then assign the same two values to our traffic generation pair that generated that traffic in the laboratory. While we consider this method to be the best practice, in this section we explore other options that are commonly practiced in the research today. Experimenters typically assign the same window size to all the connections for traffic generation in a given experiment.

What role does window size assignment play on the performance metrics in an experiment? To answer this question, we ran some experiments varying window sizes while keeping all other variables the same. But before we present the results for varying window sizes, let us briefly review the role of the receiver maximum window allocation in TCP. During the setup, or the three-way handshake, for a TCP connection, both endpoints of the connection *advertise* their *receiver maximum window size*. This is the size of the operating system buffer where the received TCP payload is stored until it can be read by the application. The endpoints communicate their receiver window size to each other to avoid any buffer overflow and resulting loss of TCP data at the receiver. The sender controls the number of unacknowledged TCP segments in the network so that this buffer does not overflow on the receiver's end. This mechanism is called *flow control* and imposes a limit on the maximum throughput of a TCP connection.

In this section, we present results for experiments using the *a-t-b-t* connection structure model with the *usernet* RTT emulation. We varied the receiver window model among these experiments to study the effect of window sizes on performance metrics. Other than for the standard Tmix model, we assigned the same window size for all connections for each set of experiments discussed in this section. Hence we assigned 8KB, or 16KB, or 64KB buffers for receiver maximum window for all connections within an experiment, and to both endpoints of traffic generation for a given experiment. First, we present the results for experiments run in the *unconstrained* mode.

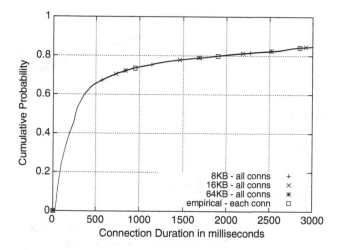

Fig. 6.30 Connection duration – CDF (a-t-b-t with usernet – unconstrained – UNC)

6.3.1 Results in Unconstrained Mode

We performed four experiments (in the *unconstrained* mode), all using the *a-t-b-t* connection structure model and *usernet* RTT model. For the first experiment, we assigned 8KB buffers as the receiver maximum window size for both sides of every one of the 4.7 million connections generated in the experiment, regardless of what the original receiver window sizes were, when we captured the traffic on the production link. For the second and third experiments we assigned 16KB and 64KB receiver window sizes for all connections at both endpoints. From here on, we refer to this receiver maximum window size as *window size*. It is not to be confused with other window sizes like congestion window which we do not manipulate directly during our experiments. For the fourth experiment, we assigned window sizes to both endpoints of every connection exactly as obtained from the original traffic data (the method employed for all experiments reported in Chapter 5).

See Figs. 6.30 and 6.31 for the CDFs and CCDFs of connection duration. We observe that there is no difference in either the body or the tail of these distributions. Figures 6.32 and 6.33 show similar results for response time distributions. That is, there is no effect of different window size assignment models on either connection duration or response times. The queue length distributions shown in Figs. 6.34 and 6.35 are as expected, since there is no constraint on the link and hence there is no queue buildup. And Fig. 6.36 shows no significant difference in the number of active connections in the network using these different window size models.

Now, why would window size have no effect on connection duration? A connection with a larger window size, say 64KB should finish faster than one with a smaller, say 8KB, window. But this is assuming that that connection has enough data to use the larger 64KB window. Hence the characteristics of the traffic being generated

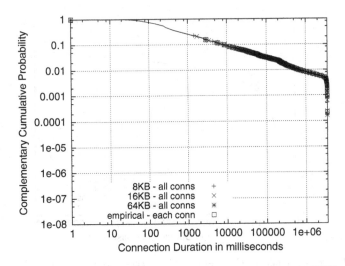

Fig. 6.31 Connection duration – CCDF (a-t-b-t with usernet – unconstrained – UNC)

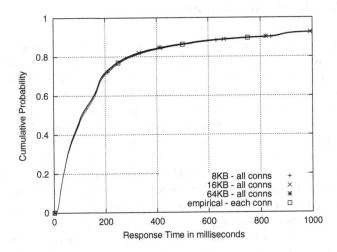

Fig. 6.32 Response Time – CDF (a-t-b-t with usernet – unconstrained – UNC)

play an influential role in this case. That is, if all connections were long and had a large number of bytes to be transferred, then of course, we would have seen significant differences with the different window size models. In that case, the 8KB windows would result in longer durations for transmitting data, while the 64KB windows would allow for much quicker data transfers and hence lead to faster completion times. On the other end of that spectrum, if all connections were short with a few hundreds of bytes to transmit, then window sizes would make no difference at all.

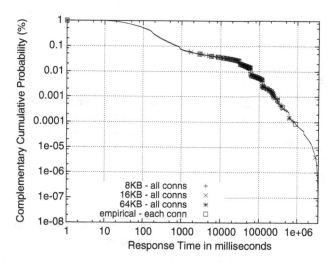

Fig. 6.33 Response Time – CCDF (a-t-b-t with usernet – unconstrained – UNC)

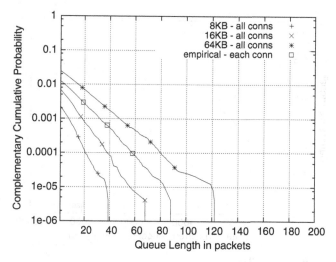

Fig. 6.34 Queue Length – CCDF (a-t-b-t connection structure with usernet RTT – different window size emulations)

Going back to some statistics on the input UNC traffic characteristics from Chapter 3 (refer Figs. 3.11, 3.13, 3.15), we note: for sequential connections, only 20% of request sizes were greater than 1000 bytes, and only 20% of response sizes were greater than 4KB. Due to a few thousand very large connections, the averages were much higher. Even so, the average request size was 2.5KB and the average response size was 11KB. For concurrent connections, the application data units (ADUs) are slightly larger than the request or response sizes of the sequential connections. Still, only 20% of ADUs are greater than 1400 bytes.

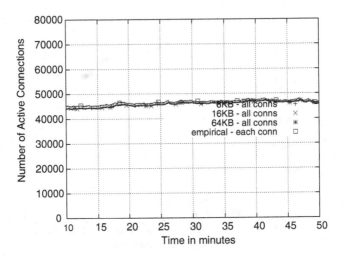

Fig. 6.35 Time series of Active connections (a-t-b-t with usernet – unconstrained – UNC)

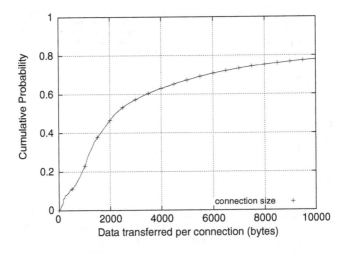

Fig. 6.36 Connection size - CDF

That was ADU or object sizes; now let us look at the connection sizes for the input data used to generate traffic in these experiments as shown in Figs. 6.36 and 6.37. Of the 4.7 million connections, 63% carry less than 4KB total data, and 37% carry more than 4KB. But only 0.6% of connections carry more than 1MB of data total; but that's still 29,000 connections. Top 10% of connections by size carry 35KB of data each – that's 470,000 connections. The mean connection size is 62KB while the median is only 2.2KB.

So, in order to observe if these larger connections within the generated traffic benefited from larger window size models, let us break down the above results in

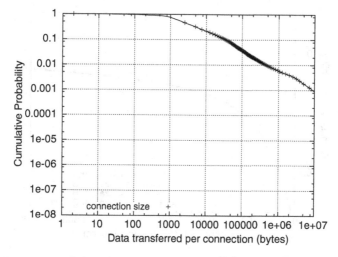

Fig. 6.37 Connection size - CCDF

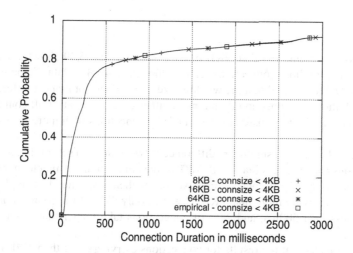

Fig. 6.38 Connection duration - CDF Connection size less than 4KB

terms of the connection sizes. We now look at the following sets of connections:
(i) all connections with less than 4KB of data to be transferred in total, (ii) all
connections with more than 4KB of data to be transferred, and (iii) all connections
with more than 1MB of data to be transferred. The third set of connections is obviously a subset of the second, but the first and second sets are exclusive sets of
connections.

In Figs. 6.38 and 6.39, we show the distribution of connection durations using all
four window size models. Figure 6.38 shows connection duration only for those

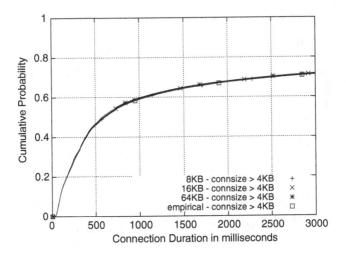

Fig. 6.39 Connection duration - CDF Connection size greater than 4KB

connections carrying less than 4KB of data, whereas Fig. 6.39 shows connection duration only for those connections carrying more than 4KB of data. Comparing the two plots above, we find that the window size model does not make a difference in the distribution of connection durations in either case – the set of all connections with less than 4KB to transfer, or the set of all connections with more than 4KB to transfer.

There is, however, a significant difference between the two sets of durations. For connections carrying less than 4KB, 78% of the connections complete in less than 500ms, and only 8% of connections take longer than 3 seconds to complete. For connections carrying more than 4KB, however, only 47% of the connections complete in less than 500ms, and 28% of connections take longer than 3 seconds to complete.

Now let us look at the results for connections carrying more than 1MB of data. The Figs. 6.40 and 6.41 show the same distributions. However, Fig. 6.40 shows durations up to the entire hour of the experimental run and indicates there is no difference in the window size models used in traffic generation. Whereas Fig. 6.41 shows durations up to only 200 seconds, and indicates that for connections carrying more than 1MB of data, if they all had 64KB buffers for receiver maximum windows, they would indeed complete much faster; for example, 40% of these connections would complete in 12 seconds or less. Using 16KB window sizes, 40% of connections would take up to 22 seconds to complete. And using 8KB, which compares with the Tmix model, 40% of connections would take up to 36 seconds to complete. Clearly, for connections carrying more than 1MB of data, window size model makes a significant difference.

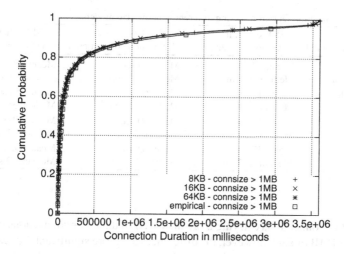

Fig. 6.40 Connection duration – CDF Connection size greater than 1MB

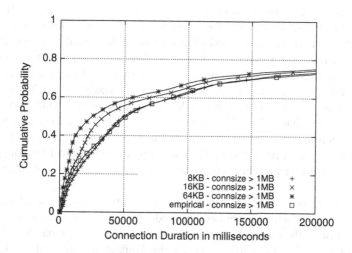

Fig. 6.41 Connection duration - CDF Connection size greater than 1MB

We now present these same results in a tabular form for a better comparison. See Table 6.1 showing the connection duration for these connections differentiated by the amount of data they had to transfer. Clearly, the mean and median duration for connections having less than 4KB of data to transfer do not change with changes in receiver window size. But the more data to be transferred in a connection, the better it can use larger window sizes. Hence we observe a slight shift in median durations

Table 6.1 Connection Duration – using different window size models

Window size model	< 4KB data transfer (3 million connections)	> 4KB data transfer (1.7 million connections)	> 1MB data transfer (28,000 connections)
8KB for all connections	Mean: 17 sec Median: 187 ms	Mean: 60 sec Median: 598 ms	Mean: 369 sec Median: 50 sec
16KB for all connections	Mean: 17 sec Median: 187 ms	Mean: 60 sec Median: 592 ms	Mean: 355 sec Median: 35 sec
64KB for all connections	Mean: 17 sec Median: 187 ms	Mean: 60 sec Median: 575 ms	Mean: 350 sec Median: 25 sec
Empirical – Tmix style	Mean: 17 sec Median: 187 ms	Mean: 62 sec Median: 603 ms	Mean: 393 sec Median: 52 sec

for the set of connections carrying more than 4KB of data. For connections carrying more than 1MB of data, however, the differences are more significant. For example, between using a 8KB window size to a 64KB window size, the median duration of these connections reduces by 50% which is a significant difference.

Comparing durations of connections for those transferring more than 1MB of data, we note that when using 8KB window sizes, the mean duration was 369 seconds, whereas when using the *usernet* model of window size assignments, the mean duration was 393 seconds. Does this mean that UNC connections with more than 1MB of data to transfer actually had windows smaller than 8KB? This seems counter-intuitive. However, when we examined the data, we found that this was indeed the case for a large number of connections. To be specific, 8.4% of connection initiators and 36% of connection acceptors had receiver maximum window sizes less than 8KB (see Fig. 3.25 in Chapter 3). Moreover, among the 28,000 connections transferring more than 1MB, 2711 connections had a window size less than 8KB for the connection initiator and 10,380 connections had a window size less than 8KB for the connection acceptor. It would be interesting to explore what kind of applications were represented by these connections, but that is currently out of scope for this study.

Now that we have seen the characteristics of the input traffic used in traffic generation, it is clearer why the window size model used in traffic generation has little to no effect on all the performance metrics when seen in aggregation for each of the performance metrics. Does this mean that the Tmix style empirical window size assignment model is too complicated and unnecessary? Before we answer that question, let us look at these same experiments run in the *constrained* mode.

6.3.2   Results in Constrained Mode

In the *constrained* mode, however, window size models seem to make a significant difference for both connection durations and response times, as shown in Figs. 6.42 through 6.45. How is that possible? Not only are there differences in the distribution

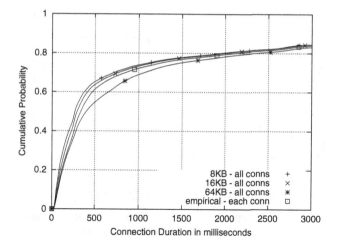

Fig. 6.42 Connection duration - CDF

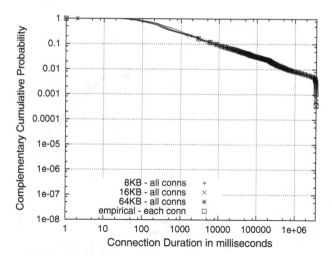

Fig. 6.43 Connection duration – CCDF

of connection durations, the results are counterintuitive. That is, the experiment in which all connections have 64KB window sizes is the one in which the connections take the longest time to complete. For example, while only 55% of connections complete in less than 500 *ms* when using 64KB windows, more than 65% of connections complete in less than 500 *ms* when using 4 KB windows.

Similarly, Figs. 6.44 and 6.45 show that connections with the large 64KB windows result in longer response times than the same connections with 4KB window sizes. The reason for the results for these two metrics lies partially in Fig. 6.46 and 6.47 which show the queue length distributions for these experiments.

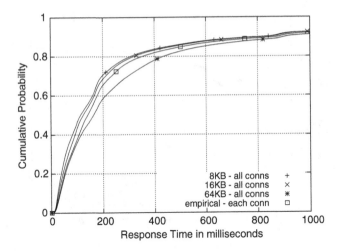

Fig. 6.44 Response Time - CDF

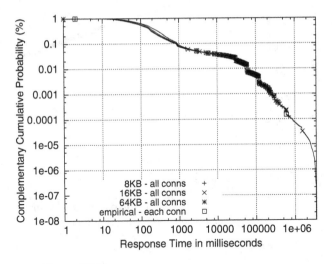

Fig. 6.45 Response Time - CCDF

The queue length distributions clearly show that the experiment with 8KB window sizes creates the lightest router queues, followed by the experiment with 16KB window sizes. The experiment with 64KB window sizes creates the heaviest queues, while the Tmix model creates slightly lighter queue lengths than the 64KB case. So, on the one hand, the connection duration and response time distributions are heaviest for the 64KB window sizes because they experience the longest queuing delays. And the connections with 8KB windows experience relatively much

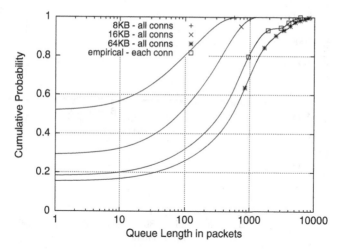

Fig. 6.46 Queue Length – CDF

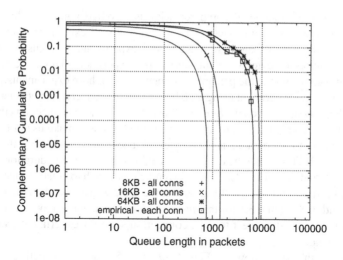

Fig. 6.47 Queue Length – CCDF

shorter queuing delays. But although this effect is due to window sizes, there is another reason.

We found that when we use 8KB or even 16KB as window sizes for all connections, there are a few thousand very large and long-lasting connections that originally had 64KB windows that are now simply unable to complete. That is they cannot send the data fast enough with these smaller window sizes. Thus the overall number of bytes transferred in these experiments with smaller window sizes is less than the original total data transferred. Hence the average link throughput is slightly lower when using

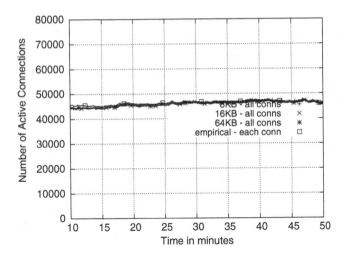

Fig. 6.48 Time series of Active connections

the 8KB or 16KB windows as compared with using 64KB windows which would obviously account for smaller queue sizes. [Note: for all experiments reported in Chapters 5 and 6 (other than these window size experiments), we ensured that the same total number of bytes were transferred per experiment. This is necessary to enable comparison among the performance metrics from those experiments.]

Now why is the experiment using all 64KB windows then creating heavier queues than the Tmix model? We conjecture that this is because those connections that have enough data to send but originally had smaller windows now make use of the larger 64KB windows to send more packets back to back, thus populating the queue. Finally, Fig. 6.48 shows that the difference in effects on connection durations among the different window size models was not significant enough to create noticeable differences in active connections; at least not at the scale we have studied them.

As we did in Section 6.3.1 for the *unconstrained* mode, let us now consider connection durations with respect to their connection sizes for experiments run in the *constrained* mode.

We start with results shown in Figs. 6.49 and 6.50 for all connections carrying less than 4KB data and more than 4KB data respectively. Clearly, for connections carrying less than 4KB of data, there is a slight difference in durations, up to 500*ms* of duration, for 8KB, 16KB, and the Tmix model of window sizes, and a very significant shift in duration for the 64KB window size model. And as we observed, this is directly due to the effect of queue lengths.

Figures 6.51 and 6.52 show connection durations for the different window size models only for connections carrying more than 1MB of data. Both figures use the same data set, with the second one zooming into the first 200 seconds of connection duration. For connections carrying more than 1MB of data, the use of 64KB window buffers is clearly helpful in completing faster, despite the effect of longer queuing delays. Why is the Tmix model showing longest duration? We conjecture

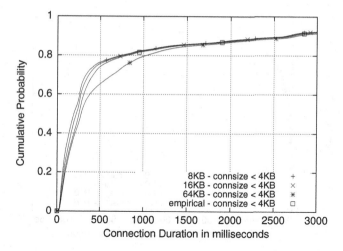

Fig. 6.49 Connection duration - CDF Connection size less than 4KB

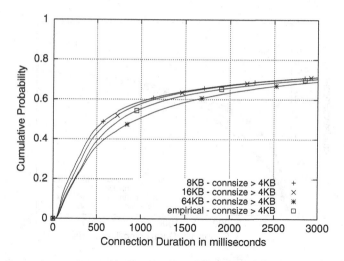

Fig. 6.50 Connection duration - CDF Connection size greater than 4KB

that this is a combination of the longer queuing delays, and possible (faithful) assignment of smaller original window sizes to these large connections.

6.3.3 Section Summary

In the *unconstrained* mode, we found that on aggregate, the window size model did not seem to make a difference in performance metrics. However, when we differentiated the connections by the amount of data they carried, we clearly saw that larger

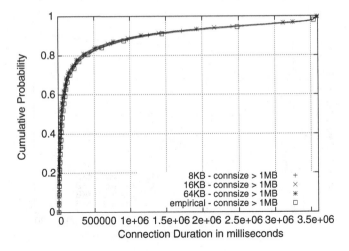

Fig. 6.51 Connection duration - CDF Connection size greater than 1MB

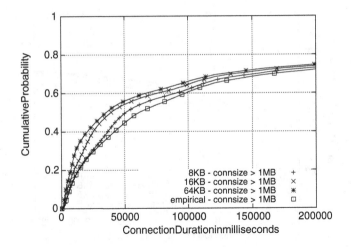

Fig. 6.52 Connection duration - CDF Connection size greater than 1MB

window sizes helped in faster completion times for connections carrying large amounts of data, for example those connections transferring more than 1MB. Then why not simply assign 64KB windows to all? While it is not a bad idea, it is also not advisable if faithful replay of traffic is a goal. That is, the pattern of injection of packets into the network for a given connection is dependent on its window size. Keeping its window size the same as was seen in the original traffic retains the packet arrival pattern into the network.

The case against using the smaller 8KB or 16KB windows for all connections is clearly laid out by the fact that large, and long-lasting connections with originally large window sizes now do not even send all the data if assigned these smaller windows. Hence, we conclude that faithfully replaying all traffic as captured clearly calls for assigning originally measured window sizes to every connection during traffic generation in the laboratory.

6.4   Connection Structure and Packet Arrival

Why does the connection structure model matter so much in traffic generation? One perspective may lie in how the structure of the generated TCP connections changes the characteristic of the overall traffic being generated in the testbed. That is, the TCP connection structure changes the pattern of packet arrivals aggregating before the core router. To study this, we observe the packet throughput in the context of the effect on queue dynamics. Specifically, we look at how different connection structures create different arrival patterns and hence create the very different queue dynamics that we saw in Chapter 5.

We study only packet arrival (and not byte arrival) patterns in this section because the queue length is measured in packets. We study the arrival of packets in the *unconstrained* mode – this gives us a sense of the traffic characteristics without the second order effect created on the traffic by the router queue in the *constrained* mode. In order to study the effect of changing connection structures, we keep the RTT and window size models the same in all these experiments; that is, we use the *usernet* RTT model and the Tmix window size model.

In Figs. 6.53 through 6.56, we show the packet throughput time series for the middle 40 minutes of each experiment, using the *block-concurrent, block-sequential, a-b*, and *a-t-b-t* models of connection structure respectively. This is the throughput as measured on the 10Gbps link before the first router on the path with higher

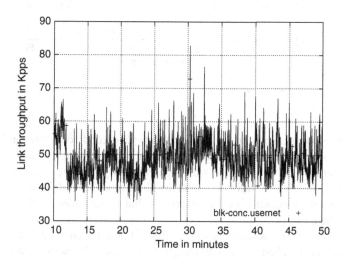

Fig. 6.53 Link throughput in packets – *blk-conc*

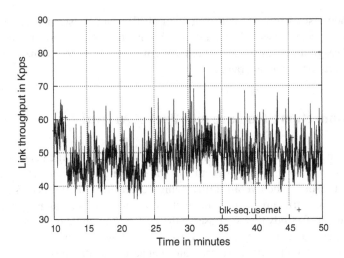

Fig. 6.54 Link throughput in packets – *blk-seq*

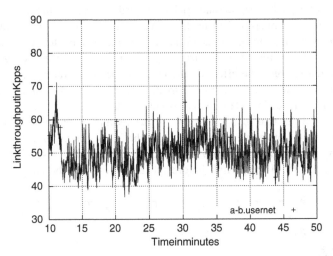

Fig. 6.55 Link throughput in packets – a-b

throughput, and is measured in *Kpps* (kilo packets per second), computed in 1 second intervals. A mere visual observation indicates that the two block models create very similar patterns of packet arrivals, with high variability, into the network. The *a-b* model shows slightly less variability than the block models, while the *a-t-b-t* model shows even less variability in packet arrivals but higher average throughput of packets in the network.

We now observe the same data shown above, but in the form of distribution of packet throughput, called throughput marginals. In Figs. 6.57 and 6.58, we show the link throughput of packet arrivals to the network in *Kpps*, in 1 second intervals and

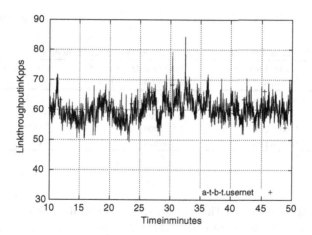

Fig. 6.56 Link throughput in packets – a-t-b-t

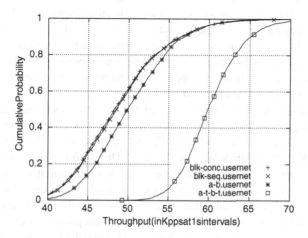

Fig. 6.57 Link throughput (packets) – 1s intervals

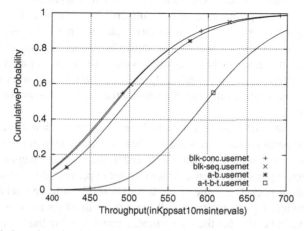

Fig. 6.58 Link throughput (packets) – 10ms intervals

Table 6.2 Packet throughput – using different connection structure models

Connection structure	Mean throughput in Kpps	Standard deviation of throughput in Kpps	Coefficient of Variation (CoV) = std_dev/mean
blk-conc	48.8 Kpps	5.6 Kpps	0.1147
blk-seq	49.0 Kpps	5.5 Kpps	0.1126
a-b	50.3 Kpps	5.0 Kpps	0.0999
a-t-b-t	60.2 Kpps	3.8 Kpps	0.0625

$10ms$ intervals respectively. This is for experiments in the *unconstrained* mode. We quantify these observations in the Table 6.2 below.

From the table, we observe that using either of the two block structures to generate the same input traffic creates slightly less packet throughput than when using the *a-b* model. The block structures produce mean packet throughputs of 48.8 *Kpps* and 49.0 *Kpps*, whereas the *a-b* model produces 50.3 *Kpps*. This is because the *a-b* model sends data in epochs in a request-response exchange pattern. This creates slightly more packets whereas in the block structure models, the same connection data is sent all at once instead of in application data units, and hence the block structures maximize packet sizes for sending the data.

The *a-b* model, however, models concurrent connections (which carry a very large percentage of bytes) exactly as modeled in the block-concurrent connection structure. This partly explains why there is not an even greater difference between the *a-b* model and the block structure models. The *a-t-b-t* model creates a major shift in packet arrival patterns. First the mean packet throughput is much higher at 60.2 *Kpps*. This is due to data being sent in application data units for concurrent connections which carry a substantially large amount of data. The combination of introducing ADUs for concurrent connections and endpoint latencies in between these ADUs results in a greater number of packets carrying the same data. The few thousand very long concurrent connections with large ADUs and several long endpoint latencies make a huge difference in the packet arrival patterns in this model.

If we look at mean packet throughput numbers alone, we would have expected the *a-t-b-t* connection structure to cause more queuing in a *constrained* environment. But that is clearly not the case as shown in results in Chapter 5 (see Fig. 5.4.53). Instead, the much higher standard deviation of the packet arrivals in small intervals for the block structures and fairly high standard deviation for *a-b* explain why these structures create different dynamics at the router queue as compared with the *a-t-b-t* model.

Given an input traffic for experimental use, the connection structure of the generated TCP connections affects the number of packets used for the same data. That is, for the same amount of data transferred, if there are more epochs in a connection transferring this data, then a higher number of packets are generated for that connection. More packets for the same data also result in more packet overhead which then slightly increases the overall throughput in Mbps as well. Intuitively, this would have indicated that the *a-t-b-t* model would result in the heaviest queues.

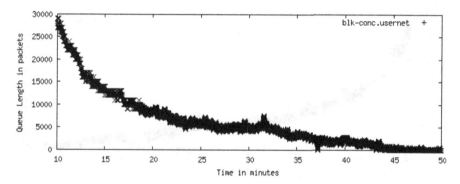

Fig. 6.59 Queue Length Time Series – *blk-conc*

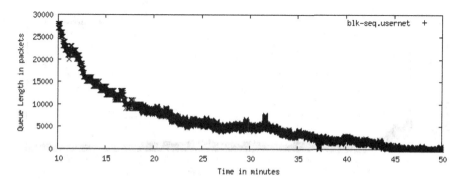

Fig. 6.60 Queue Length Time Series – *blk-seq*

But the experimental results show counter-intuitive queue dynamics. Higher average throughput in packets should result in larger queues, right? Wrong, or not always. It's just not that simple. When these higher average throughput for packets are accompanied by endpoint latencies, such as in the *a-t-b-t* model, this allows time for the queue to drain, and hence the queues are smaller on average.

A major problem with not using endpoint latencies within connections is that large connections send packets at a faster pace occupying the queue, which creates queue buildup that takes an inordinate amount of time to drain. Of course, one could argue that we used an unrealistic queue size – but, as explained in Chapter 4 on methodology, that is besides the point here. To explain this queue buildup, we show in Figs. 6.59 through 6.62 the queue length time series of the router queue when the router-to-router link was in the *constrained* mode.

These figures show the queue length time series for only the middle 40 minutes of the experiments. Even so, as seen in Figs. 6.59, 6.60, and 6.61, the initial queue buildup is so drastic that the queue has not drained well into more than half hour in to the experiment. Thus, using the block structures and even the *a-b* model for connection structure builds up the queue very quickly, and here's why. Large, long connections that start at the beginning of the experiment, and would have lasted for, say,

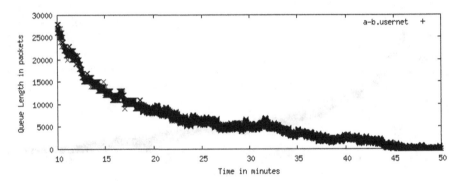

Fig. 6.61   Queue Length Time Series – a-b

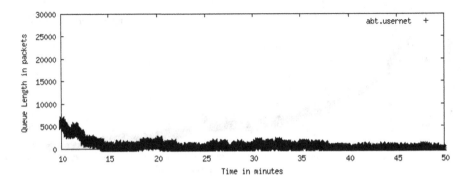

Fig. 6.62   Queue Length Time Series – a-t-b-t

3 minutes or even up to 40 minutes into the experiment, when replayed without any of the endpoint latencies, now send packets into the network at a much faster pace. These connections build the queue quickly such that it takes a long time for the queue to recover from this buildup. Sure, a shorter queue would have alleviated this dynamic, but we wished to eliminate the loss dynamics that would result from a shorter queue, and study the effect on traffic characteristics due to the different connection structures. When the *a-t-b-t* connection structure model is used, as shown in Fig. 6.62, replaying the endpoint latencies as was present in the original traffic allows the queue to drain. Thus even though there is an initial queue buildup, the queue is able to drain by just 12 minutes into the experiment. This does also bring up the question about what constitutes an appropriate time for running an experiment. We discuss this among other such methodological questions in Chapter 7.

   We conjecture that the differences in round trip time emulation schemes created similar effects as shown above for the different connection structure models. That is, larger average RTTs for connections means more time between packets for the queue to drain, and hence less average queue lengths as we observed in Chapter 5 (see Figs. 5.57 through 5.64).

6.5 Long Range Dependence

Router queue dynamics are affected by many factors. One of them is the long-range dependence (LRD) characteristic of the traffic. In this section, we explore whether changing the connection structure model changed the long-range dependence of the traffic by studying the wavelet spectrum and computing the Hurst parameters (see Table 6.3) for these packet arrivals.

Figures 6.63 through 6.66 show the wavelet spectrum for packet arrivals for the *block-concurrent, block-sequential, a-b* and *a-t-b-t* models, all using the *usernet* RTT model for traffic generation, using the same input UNC traffic in all these experiments.

We found that the long-range dependence characteristics of the packet arrival time series remains the same regardless of the connection structure used for traffic generation. How is that possible? While not all the factors affecting LRD have been clearly identified in networking research studies, we know that the distribution of connection sizes, distribution of object sizes, connection arrival times, and the distribution of round trip times all play an important role in the LRD characteristic of traffic.

Changing the connection structure only changes the inter-packet arrival times within a connection. For example, lack of epoch structure and endpoint latencies results in packets being sent back to back more frequently in the block structure models. That is, we keep the connection sizes the same in all these models, but the

Table 6.3 Estimated Hurst parameters and their confidence intervals for packet throughput time series using the four different connection structure models

Connection Structure	Hurst parameter	Confidence interval (95%)
blk-conc	0.9636	[0.92759, 0.9996]
blk-seq	0.9659	[0.92991, 1.0019]
a-b	0.9826	[0.92548, 1.0396]
a-t-b-t	0.9631	[0.906, 1.0202]

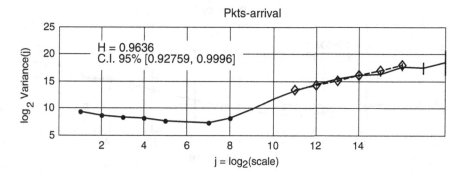

Fig. 6.63 Wavelet spectrum for packet throughput time series using the block-concurrent connection structure model

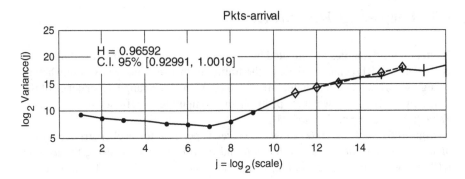

Fig. 6.64 Wavelet spectrum for packet throughput time series using the block-sequential connection structure model

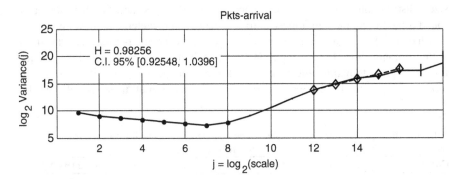

Fig. 6.65 Wavelet spectrum for packet throughput time series using the a-b connection structure model

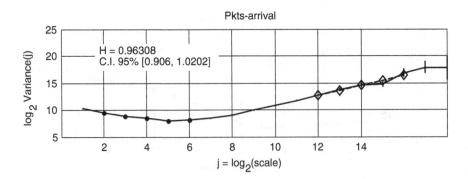

Fig. 6.66 Wavelet spectrum for packet throughput time series using the a-t-b-t connection structure model

two block structure models change all object sizes within each connection. However, we leave the other factors unchanged. Connection sizes, start times of connections, and round trip times were all retained as part of the traffic generation discussed above, even as we changed the TCP connection structure model for generating these connections. Hence, changing the connection structure model alone did not affect the LRD characteristics of the traffic generated on the link in the laboratory testbed.

6.6   Chapter Summary

In this chapter, we presented results for experiments using four RTT models not discussed in Chapter 5. The results for all four performance metrics for these RTT models support and supplement the results seen for the three main RTT models presented in Chapter 5. Of these, the *DA* RTT model emerged as a close approximation of the more realistic *usernet* RTT model. That is, using the *DA* model for RTT emulation in traffic generation is an acceptable alternative to the *usernet* RTT model.

We also presented results for varying the receiver window size model for traffic generation. While window size assignment does not seem to affect most of the TCP connections, the window size model makes a large impact on performance metrics for connections carrying more than 1MB of data. From the results obtained in that section, we recommend using the Tmix model of window size assignments. Finally, we discuss how changing the connection structure model changes the pattern of packet arrivals into the network and hence causes changes in performance metrics.

Chapter 7
Conclusion

> *The most exciting phrase to hear in science, the one*
> *that heralds the most discoveries, is not "Eureka!",*
> *but "That's funny..."*
>
> ISAAC ASIMOV

> *There are two possible outcomes: if the result confirms*
> *the hypothesis, then you've made a measurement. If the result*
> *is contrary to the hypothesis, then you've made a discovery.*
>
> ENRICO FERMI

This study was a methodological study seeking some fundamental insights into experimental methods in networking. Specifically, we looked at methods and modeling for traffic generation in empirical networking research. We plan to use the lessons learned from this study to motivate further discussions and take concrete steps to engage the networking research community toward establishing better practices in experimental methods for networking research.

We examined the effects of several choices for modeling network traffic generated for empirical research on the following application-level and network-level performance metrics: connection duration, response time, router queue length, and number of active connections in the network. We examined the choices in modeling TCP connection structure, connection round trip times (RTTs), and receiver window sizes within the realm of realistic traffic generation. In this chapter, we discuss our findings, their implications, and some related work we wish to explore in the future.

J. Aikat et al., *The Effects of Traffic Structure on Application and Network Performance*, 291
DOI 10.1007/978-1-4614-1848-1_7, © Springer Science+Business Media New York 2013

7.1   Observations and Conclusions

To arrive at the conclusions presented in this chapter, we conducted extensive sets of experiments using the Tmix traffic generation system on a 10Gbps laboratory test-bed, using four different connection structure models, seven different RTT models and four window size models. We ran our experiments using two very different traffic inputs and in two different network environments. Here are our key observations.

> In an unconstrained network, regardless of the connection structure model used, or the input traffic used, round trip time had a significant effect on application-level performance measures of connection duration and response times, but only up to a maximum of 1 second for these metrics.

Consider this example. You are designing a set of experiments to evaluate a new Active Queue Management (AQM) protocol. This is a study in which application-level performance metrics are important. In fact, your study relies on response time as the main metric that distinguishes whether this new AQM scheme is better than, say, the currently used drop-tail method of queue management. Which RTT model will you pick for generating traffic? If you pick the *meanrtt* model, or any similar model that emulates a single path for all the millions of connections traversing the router that has implemented the new AQM scheme, are you creating a realistic scenario for testing that AQM scheme? When it is deployed on routers in the middle of the network, those routers will service connections with a wide range of RTTs each possibly traversing completely different paths through the Internet.

Say, you obtain results for this AQM scheme using the *meanrtt* model, and find that it performs worse than the drop-tail scheme. Could you then be confident that your AQM scheme was not good enough for deployment? We assume you picked the *meanrtt* value from the empirical mean of the original RTT distribution. Still, is it possible that the artificial emulation of one single path for all connections created queue dynamics that would have been different if there had been a wide range of realistically possible connection RTT values? So, then let's say you picked the *usernet* RTT model instead. Could you now be sure of the results and use it for deployment? At least, you could be sure in this case that the characteristics of traffic you used is as similar as possible to real network traffic, and hence the results obtained would be much more reliable.

Of course, network traffic keeps changing and is different at different points on the Internet. Hence it would be even better if you could run your entire set of experiments using two or more input traces with significantly different application mixes. Why? Traffic characteristics play a huge role in the outcome of experiments as we saw in this study. The UNC and IBM traffic had many similar characteristics when seen on a high level, but there were significant differences. For example the average RTT for UNC connections was 80 ms and for IBM connections, it was 92 ms. If the results you obtain from your AQM study hold for two such very different traffic mixes, then that in itself will serve to make your results more reliable when you make the argument for deployment of your new AQM scheme.

> RTT model had no impact on the number of active connections in the network.

This result must be qualified with the following statement: we measured active connections for every second in the hour long experiment. In fact, since connection durations affect the number of active connections, and RTT models had no effect beyond 1 second for connection durations, it is expected that RTT models would have no effect on active connections. So, in this case, say your experiment involved testing a routing protocol that had to keep state for active connections traversing that router queue. What is the granularity at which you wish to update and compute the number for active connections? If it is every second, then the RTT model you use will not greatly affect this computation. But if it is every millisecond, then our result above would have changed. That is, the RTT model would have affected your experiment. Hence, it is important to verify the scale at which your choice of traffic generation (RTT model in this case) will or will not affect your metric of performance.

> In a constrained environment, the lighter the distribution of connection RTTs, the heavier was the queue distribution at the router.

Choosing the right RTT model for your study depends not only on the direct effect of the RTT model on application-level performance metrics but also on the indirect or second-order effect of queue lengths on such application-level metrics. For example, in our study, we deliberately chose a very high level of congestion, which was induced by constraining the router-to-router link at 105% of the traffic on that link. This created long queues at the router before this link, which added very large queuing delays to connection durations and response times. If your study did not involve *constrained* links and created no heavy congestion scenarios, then perhaps you would not have to take into effect these second-order queuing effects on your application-level metrics of performance.

> Randomly assigning the same empirically derived round trip times to connections, using the discrete-approximation (DA) RTT model, is almost as effective, on an aggregate level, as assigning each connection its originally measured RTT as done in the usernet model.

We developed a model for RTT emulation that approximated the empirical RTT distribution and emulated 30 paths for connection RTTs in the experiment. We concluded, from our experimental results, that this model would mimic the *usernet* model quite closely for the metrics of performance used in our study. We recommend that this is a viable alternative to the *usernet* RTT model in cases where the *usernet* model cannot be deployed. However, we found that if queuing dynamics in highly congested environments is of interest in an experiment, then this model is not appropriate since it produces heavier queues than the *usernet* model. We used 30 values to approximate the 4.7 million connection RTTs in our traffic. We conjecture that if more paths were used, for example 150 paths were used, then this *DA* model would produce results that would even more closely resemble the *usernet* model. The choice of the number of paths for this model was purely dependent on the topology of the physical laboratory network in our study.

> The impact of the RTT model used in traffic generation, while significant, becomes negligible when compared to the dramatic impact of the connection structure model used in the experiment.

We strongly recommend that experimenters design their methodology carefully and select an appropriate RTT model for their study for all the reasons already stated above. We are convinced from the results in this study, however, that while modeling connection RTTs is still a choice that depends on the overall experimental design and goals, there is less of a choice in picking a connection structure model. But did we not show four choices for connection structure models in this study? Yes. However, we have come to the following conclusion that restricts the choice of connection structure model. That is, both the size and time components are just as important in modeling connection structure. Simply modeling connections by their size (as we did in the two block structures), or even adding the number of objects and size of objects (as we did in the a-b model), or further preserving the request-response exchanges (as we did in the a-b model) are not enough detail in connection structure modeling. The one component of endpoint latencies within connections creates such a dramatic effect on all metrics of performance (both application-level and network-level) that we are convinced that connections must be modeled at this level of detail; that is we must include the epoch structure as well as all endpoint latencies.

Is the a-t-b-t model then the only correct model for traffic generation? It is definitely one method of detailed connection structure modeling, and it is the one we explored in this study. However, there may be others that work just as well, but were out of scope in this study. For example, how would the results have differed if we used the number of epochs per connection, the epoch sizes, and the endpoint latencies as input distributions to our traffic generation system? In such a case, we are indeed including details of connection structure but have not preserved the correlation, if any, among these various components of connection structure within connections. The a-t-b-t is a non-parametric model while what we just described here is a parametric model for connection structure, similar to the Swing traffic model [1]. Is one better than the other, or more realistic than the other? Would these two models produce similar or very different results? That is, given the same input traffic, let's say we use Swing for parametric modeling and Tmix for non-parametric modeling to generate traffic. Thus with inputs from the same empirical measurements, it would be interesting to bring out the similarities and differences, strengths and weaknesses of the two modeling techniques. These parametric distributions usually represent millions of connections; at such high levels of aggregation, how does parametric modeling compare with its non-parametric counterpart? These are all open research questions and intended for future work.

Unlike RTT models which affected connection duration and response times only up to 1 second, the connection structure models affect these metrics significantly in the body as well as the tail of the distribution for these metrics. Hence, the connection structure model greatly affects the number of active connections in the network as well. And in the constrained mode, the absence of endpoint latencies in the block structures and the a-b model result in much heavier queues at the router, thus creating counter-intuitively long durations and response times because of the second order effects of queuing delay on connection duration and response times.

As we mentioned above, the connection structure model affects application-level metrics very significantly and throughout the experiment duration. Hence, if we used our previous example of evaluating a new AQM scheme in a router, choosing a realistic model for connection structure becomes very important for a reliable study. If we eliminated endpoint latencies from the model, there is a multi-fold decrease in the number of active connections in the network. And in *constrained* mode, the queuing dynamics would be very different for the different connections structure models. For example, say you used one of the block structure models in your study and determined that your protocol could keep state for active connections for a certain level of network traffic. Now, in the real network, the traffic resembles the *a-t-b-t* model where the number of active connections is multi-fold that for block structures, and your protocol may fail in this scenario.

It must be noted that in order to isolate and study queuing dynamics, we deliberately set the router queue to 65,000 packets for all our experiments. While this might be considered unrealistic, this helped us study the queue dynamics in the absence of loss within TCP connections. For our experiments in the *constrained* mode, the connection structure model used for traffic generation had dramatically different queue dynamics at the outbound queue of the router. This caused very large queuing delays for these connection structures and hence created a second order effect on all other performance metrics due to the queuing delay. Sure, if we had shorter queues, say 1200 packets, these connections would have incurred losses and created a different dynamic in the network. But we designed the queue size with the intention of eliminating losses since study of loss characteristics was not part of our goals. This is definitely a topic that we intend to explore in the near future. That is, set the queue size to different levels inducing loss and study how this changes the traffic characteristics and the effect it has on the different performance metrics at both application and network levels.

The choice of this very large queue size at the router caused the block structures and the *a-b* model to buildup long queues at the beginning of the experiment, and these long queues did not drain until well into more than half the experiment duration of one hour. This leads us to another open question for experimenters: what is an appropriate length of time to run an experiment? It may be different depending on traffic characteristics of the input trace being emulated, as well as on network characteristics during the experiment. However, there seems to be no consensus on this, except to say that you must have a *stable* region from which to derive results. Such a stable region remains to be clearly defined. Five minutes for an experiment seems to be an acceptable time for running experiments, and used in some leading papers. For example, Swing [1] uses very small traces of five minutes and up to a maximum of twenty-two minutes. The open question here is: what constitutes stability in an experiment? Is it that the input must attain stability? For example, in our experiments, we see a spike in the throughput in the middle of the network as all 30 pairs of traffic generators start up. It takes about 5 minutes for the throughput to settle down, and hence we use 10 minutes into the experiment as the start of our *stable* region. This stabilizes the input. But is stability of the experiment defined by the effect on performance metrics? For example, should we wait until the router queue stabilizes? Is that really achieving stability in the experiment, or ignoring the effects of traffic generation models?

> The take away message, if there is to be just one, is that the time components of traffic generation are as important as the size components.

We simply wish to emphasize that experimenters must take into account the endpoint latencies when designing a model for connection structures in their experiments for all the reasons already enumerated above.

> For the bulk of connections in any experiment, window size assignment made no difference in connection durations or response times.

We found that the window size assignment model for assigning receiver maximum windows does not seem to affect the bulk of connections in our experiments, when run in the *unconstrained* mode. This is because the bulk of connections carry a small number of bytes and last for a short time. However, we also found that the window size makes a huge difference in these metrics for connections transferring more than 1 MB of data. While this is not surprising, it is noteworthy that these usually small number of connections carry a relatively large percentage of the bytes, and hence they do affect network-centric metrics like queue length in a congested environment and number of active connections in the network. Moreover, the queuing dynamics of using a single value of window size for all connections in an experiment changes significantly in a highly congested network environment. For this reason, and for preserving the network-level pattern of injection of packets for large connections, we recommend using the Tmix model of window size assignment for traffic generation. It would also be interesting to measure how often window scaling is used in real connections. We did not measure or try to emulate window scaling in our study.

> Changing the connection structure model (to the extent done in this study) for a given input traffic does not change the long-range dependence characteristic of the packet arrival time series generated using these different connection structures.

We found that the long-range dependence characteristics of the packet arrival time series remains the same regardless of the connection structure used for traffic generation. This is because while we removed endpoint latencies and even epoch structure in some models, we retained connection sizes and round trip times. It would be interesting to study which components of connection structure affect LRD of the traffic generated. For example, for a given set of connection sizes that cause LRD in traffic, is it the large connection sizes or the feedback/pacing of TCP that causes LRD? That is, exactly what components add LRD characteristics to traffic? And how does varying the LRD affect the various metrics in this study? It would be interesting to experiment with the following designs to study their effect on the LRD of the generated traffic: inter-packet times at the IP level, packet arrivals within TCP flows, flow arrivals at the TCP level, inter-arrival time for start times of TCP flows, use only top 30% of flows by duration, or use only top 30% of flows by bytes.

> The outcome of any experimental evaluation depends heavily on the input to the system – this is the garbage-in garbage-out concept.

The more realistic the generation of traffic, the more reliable the outcome will be of the empirical research. Hence, if we wish to run experiments with the goal of evaluating a new or improved network protocol, then we should test this protocol using realistic network traffic. Of course, there is no *standard* network traffic. Indeed traffic captured at one location on the Internet could be vastly different from any other location on the Internet. This is why it is desirable to use input traffic captured at a production link and preferably at more than one such production links. Reproduction of traffic on a link should include all the traffic on that link. However, in this study and in most others, we only consider TCP traffic which constitutes over 90% of all traffic on the Internet. How would UDP traffic generation affect the overall results obtained in this study? Although non-TCP transport protocols, mainly UDP, are a small fraction of the traffic on the Internet today, they are almost always left out of traffic generation systems. It would be useful to have UDP traffic as one of the suite of traffic scenarios in an experimental standards suite.

7.2 Modeling Traffic

A major goal of traffic generation on the experimental network is to represent the original mix of applications while doing so without knowledge of what the original applications were. Changing the connection structure model, however, effectively changes the application mix in the original traffic by changing the behavior of the application as well as the behavior of the end user in some cases. This causes TCP to send packets in a different pattern, in both size of packets as well as the time elapsed between successive packet transmissions. This change at the TCP connection level due to change in application behavior gets amplified when playing tens of thousands of connections simultaneously, and this alters the aggregate arrival pattern of the traffic to the network link.

Similarly, changing the round trip time (RTT) for each connection, while keeping the connection structure model and hence application behavior unchanged, changes the pace at which each individual TCP connection sends windows of packets into the network. The dynamics of the TCP feedback loop are heavily influenced by the RTT for the connection. Smaller the RTT faster is the feedback from one end of the TCP connection to the other, leading to a quicker growth of the congestion window for that connection. This means that a connection with smaller RTT results in quicker transmission of the same application data. And similarly, a larger RTT results in slower feedback, slower transmission of data and larger completion times for connections. So, RTT plays a role in both propagation time for the TCP packets and also the time taken for the window to grow and allow for faster transmission of data. When such a change is effected at the level of every TCP connection, the aggregate traffic resulting from this change creates a different pattern of arrival of packets to the network. Window size changes in TCP connections similarly affect the growth of the window and thus the amount of unacknowledged packets in the network for a given TCP connection. Hence larger window size means the TCP connection can transmit the same data faster and have more data in the network before it receives feedback from the other end of the connection.

If connection structure modeling, RTT emulation methods and window size assignment each have such significant impact on every TCP connection, the expectation would be that every one of these input changes would see very drastic changes in the traffic characteristics of the resulting input traffic to the network link. However, that is not the case. The changes are more pronounced in some cases than others. This is largely due to the fact that today's Internet links constitute load from a very large number of connections, most of which are small in size in terms of the bytes they transfer. A significant percentage of the connections are not large enough to take advantage, or be adversely affected as the case may be, of the changes in RTT or window size, and in some cases, of the changes in connection structure models as well.

The importance of changing these input variables is, however, significant when we consider that they have a considerable impact on large connections. Though such large connections constitute a small fraction of the number of connections in the traffic on any given Internet link, they tend to carry a disproportionately large amount of bytes and packets on the link, and thus contribute heavily to the overall characteristics of traffic on that link.

7.3 Chapter Summary

In this chapter, we discuss the main observations and conclusions reached in this study. We made some recommendations for experimenters to consider as they design experiments and model traffic for networking research. In the long term, the networking research community needs some clearly defined and accepted standards for testing protocols, one that is a suite of tests that is maintained and constantly updated by the research community. This suite would contain various types of emulation scenarios with various types of input, and measurement tools for studying various performance metrics. Using such a testing suite, a researcher proposing a new protocol or an improvement to an existing protocol could clearly show that it would improve performance for specific metrics using different traffic mixes.

Reference

1. Vishwanath KV, Vahdat A (2009) Swing: Realistic and responsive network traffic generation. IEEE/ACM Transactions on Networking